MULTIMEDIA ONTOLOGY
REPRESENTATION AND APPLICATIONS

MULTIMEDIA ONTOLOGY
REPRESENTATION AND APPLICATIONS

Santanu Chaudhury
Indian Institute of Technology, Delhi

Anupama Mallik
Indian Institute of Technology, Delhi

Hiranmay Ghosh
TCS Research, Delhi

CRC Press
Taylor & Francis Group
Boca Raton   London   New York

CRC Press is an imprint of the
Taylor & Francis Group, an **informa** business

A CHAPMAN & HALL BOOK

The cover image features the famous Dragon's Blood Tree "Dracaena Cinnabara," found on the Socotra island in the Indian ocean. This island separated from its continent six million years ago. This ancient tree species has a unique appearance with its umpteen branches emanating from its trunk, and its leaves which sprout only at the end of its youngest branches. The Socotra dragon tree epitomizes an ontology with an infinite amount of knowledge embedded in its innumerous linkages and connections. Ontologies can help generate new knowledge in just the way new Socotra dragon trees are shown to take root in surrounding areas. Multiple hues of color in nature represent the media aspects of a multimedia ontology knowledge-base.

CRC Press
Taylor & Francis Group
6000 Broken Sound Parkway NW, Suite 300
Boca Raton, FL 33487-2742

© 2016 by Taylor & Francis Group, LLC
CRC Press is an imprint of Taylor & Francis Group, an Informa business

No claim to original U.S. Government works

Printed on acid-free paper
Version Date: 20150521

International Standard Book Number-13: 978-1-4822-3634-7 (Hardback)

Visit the Taylor & Francis Web site at
http://www.taylorandfrancis.com

and the CRC Press Web site at
http://www.crcpress.com

To Our Teachers ...

Contents

Foreword

The 21st century started a major transformation in knowledge creation, storage, and sharing. Gutenberg's printing technology brought the era of text documents that dominated human knowledge dynamics for about six centuries. Technology once again started transforming knowledge mechanisms. Mobile phones and their cameras, good bandwidth, increased storage and processing power are bringing a major unprecedented revolution in society. Photos and videos have started becoming the new documents. Until the last century, documents were textual with the use of photos and figures to enhance them. Now documents are becoming visual, with the use of text to supplement them. After about two decades of bringing rapid advances to society, the World Wide Web started becoming increasingly visual and is likely to see the emergence of the Visual Web; the web that will be formed by linking photos using implicit as well as explicit links. I am fascinated and intrigued by this transformation.

And then comes this book — Multimedia Ontology: Representation and Applications. What a timely arrival for the book! Ontology was commonly used to represent textual knowledge. Of course textual knowledge will always remain a key component of any knowledge representation because humans developed a symbolic structure of knowledge based on the semantics of the real world using language. And text is a visual representation of that language. But as our technology allowed us to capture visual and other sensory knowledge, suddenly traditional ontology started appearing limited. Extensions to ontology for capturing this multimedia data and information appeared essential for the advancement of our knowledge ecosystem.

This book addresses the important need for developing efficient technologies to manage multimedia content. Traditional ontology representation uses linguistic structures for representing concepts and relations in a domain. This is very limited in efforts to interpret multimedia data, which is an experiential record of the world. In their book, Chaudhury, Mallik, and Ghosh establish the requirement for an ontology representation scheme for multimedia applications. Such ontology should be able to represent a domain not only in terms of concepts and relations, but also in terms of the media properties of concepts and the relations between them. In an emerging multimedia knowledge world, ontology should be able to deal with a unified description of conceptual and experiential worlds.

This book reviews the state of art in representing multimedia ontology and their use in several applications. The authors analyze the issues and bottlenecks in existing approaches. To overcome the limitations in the existing approaches, they develop a new ontology representation scheme and define a representation language called "Multimedia Web Ontology Language" (MOWL). MOWL provides a mechanism for encoding the media properties of concepts and to reason with them in a domain context. This is a strong contribution of this book that is likely to be a long-lasting contribution to multimedia knowledge representation.

The second half of this book discusses how MOWL can be used in different application scenarios. For new applications, ontology must be created. This book discusses how the ontology can be created using different means, such as by eliciting knowledge from domain experts and by analyzing and adapting the traditional knowledge resources. They develop a reasoning approach using MOWL, which can be used for constructing holistic observation models for concepts that manifest as complex media phenomena. The observation models can be used for robust recognition of abstract concepts, which is not possible with other techniques. By considering different applications, this book presents a complete cycle of ontology creation, reasoning using ontology, and recognition of concepts and actions in multimedia data.

In the rapidly evolving world of changing documents where photos and videos are likely to be dominant, approaches to represent knowledge effectively are essential. This book is the first book addressing this problem and it comes at a very appropriate time. I was delighted to see this book.

Prof. Ramesh Jain,
Bren Professor in Information & Computer Sciences,
Donald Bren School of Information and Computer Sciences,
University of California, Irvine.
$http://ngs.ics.uci.edu$

Preface

Proliferation of multimedia data on the web is driving the need for developing effective schemes for management of multimedia contents by exploiting embedded semantics of the multimedia data. However, in this context, the key challenge is that of designing an appropriate representation scheme for multimedia knowledge so that content can be interpreted and used for a variety of information processing tasks. Ontology and ontology languages provide a framework for semantically exploiting numeric and textual content on the web. Multimedia ontology can enable development of a similar capability, but it has to be distinct in terms of the structure, syntax and inference scheme because of the unique nature of the multimedia content.

In this book, we provide comprehensive coverage of different aspects of multimedia ontology and its applications. We examine issues involved in multimedia ontology design and its exploitation with a new perspective in the context of different approaches available in the literature. This book deals with formal and applicative aspects of ontology engineering for multimedia content, where exposition of the relationship between content-processing schemes with formal ontology-based inference strategies have been dealt with. We have tried to present a consistent, cohesive and comprehensive account of the relevant research.

This book is targeted towards researchers and advance graduate students working in the areas of ontology engineering, multimedia information retrieval and the Semantic Web. Topics from chapters 4, 5, 6 and 7 can be used in a course on multimedia data management for covering different aspects of multimedia ontology.

This book is based upon research work undertaken by the authors and their collaborators. At different points in time, the Department of Science and Technology and Department of Electronics and Information Technology, Government of India, have supported this research effort. Support from different industrial sponsors, namely, AOL, Yahoo and TCS, is also gratefully acknowledged.

The authors would like to acknowledge help and contributions of their students and colleaugues at IIT Delhi, in particular Gaurav Harit, Ankur Jain, Manasi Matela, Brindaduti Maiti, Karthik Kashyap T., P. Poornachander, Nisha Pahal, Ronak Gupta and Deepti Goyal, and Stuti Ajmani of TCS. The authors would like to thank the team from the publishers Aastha Sharma, Jessica Vakili, Marsha Pronin and Marcus Fontaine, for their valuable input during the writing of the book. The authors would like to specially thank Medha Malhotra, graphic designer from NID Ahmedabad, India, for all her help with the figures in the book and her creative design of the cover page.

Santanu feels that this book would not have been possible without the intellectual and emotional support of his wife, Lipika. Anupama is grateful to her family for their patience with her perseverance and wishes to thank her niece Medha for her timely help with illustrations. Hiranmay thanks all his colleagues, his friends and his spouse for bearing with his erratic hours during authoring this book.

Authors

List of Figures

List of Tables

List of Acronyms

AmI	Ambient Intelligence
ASR	Automatic Speech Recognition
BN	Bayesian Network
BNF	Backus-Naur Form
BoW	Bag of Words
CCTV	Closed Circuit Television (cameras)
CIDOC	Centro Intercultural de Documentación
CIDOC CRM	CIDOC Conceptual Reference Model
CLD	Color Layout Descriptor
CNN	Convolutional Neural Network
COMM	Core Ontology for Multimedia
CORS	Cross-origin Resource Sharing
CPT	Condional Probability Table
CRM	Conceptual Reference Model
CSI	Context-specific Independence
CVB	Conceptual Video Browser
D	Descriptors (MPEG-7)
DAG	Directed Acyclic Graph
DCD	Dominant Color Descriptor
DDL	Data Description Language
DIA	Digital Item Adaptation (MPEG-21)
DL	Description Logic
DOM	Document Object Model
DS	Description Schemes (MPEG-7)
DSCT	Dual Source Computed Tomography
DTD	Document Type Definition
EHD	Edge Histogram Descriptor
E-MOWL	Event-Multimedia Web Ontology Language
FBN	Full Bayesian Network
FMA	Foundational Model of Anatomy (ontology)
FOIL	First Order Inductive Learner
GMM	Gaussian Mixture Model
GoF	Group of Frames
GoP	Group of Pictures
GPU	Graphic Processing Unit
GUI	Graphic User Interface

HDFS	Hadoop Distributed Filing System
HMM	Hidden Markov Model
HoG	Histogram of Oriented Gradients
HSV	Hue-Saturation-Value
HTD	Homogenous Texture Descriptor
HTML	HyperText Markup Language
HTTP	HyperText Transfer Protocol
ICD	Indian Classical Dance
IRI	Internationalized Resource Identifier
JSON	JavaScript Object Notation
KL	Kullback-Leibler
KNN	K-Nearest Neighbor
LDA	Latent Dirichlet Allocation
LOD	Linked Open Data
LPC	Linear Prediction Coefficients
LSCOM	Large Scale Concept Ontology for Multimedia
LSI	Latent Semantic Indexing
MDL	Minimum Description Length
MDS	Multimedia Descriptor Schemes
MFCC	Mel-Frequency Cepstral Coefficient
MOWL	Multimedia Web Ontology Language
MPEG	Moving Picture Experts Group
MVCO	Media Value Chain Ontology
N3	Notation 3 (Logic)
NLP	Natural Language Processing
NS	Name Spaces
OCR	Optical Character Recognition
OM	Observation Model
OWL	Web Ontology Language
OWL2	Web Ontology Language Version 2
pLSI	Probabilistic Latent Semantic Indexing
PODD	Phenomics Ontology Driven Data repository
QoS	Quality of Service
RDF	Resource Description Framework
RDFa	RDF (Resource Description Framework) in Attributes
RDFS	RDF (Resource Description Framework) Schema
REL	Rights Expression Language (MPEG-21)
RIF	Rule Interchange Format
SCD	Scalable Color Descriptor (MPEG-7)
SCFL	Semi-supervised Cross Feature Learning
SCH	Spatial Chromatic Histogram
SIFT	Scale Invariant Feature Transform
SMIL	Synchronized Multimedia Integration Language
SMPTE	Society of Motion Picture and Television Engineers
SPARQL	Simple Protocol and RDF Query Language

SSL	Secure Sockets Layer
SSN	Semantic Sensor Network
STFT	Short Time Fourier Transform
SVM	Support Vector Machine
SWRL	Semantic Web Rule Language
TGM	Thesaurus of Graphical Material
TLS	Transport Layer Security
UML	Unified Modeling Language
URI	Unified Resource Identifier
URL	Unified Resource Locator
W3C	World-Wide Web Consortium
WSDL	Web Service Description Language
XML	Extented Markup Language
XSD	XML (Extensible Markup Language) Schema Description

Author Biographies

Santanu Chaudhury did his B.Tech (1984) in Electronics and Electrical Communication Engineering and Ph.D (1989) in Computer Science and Engineering from Indian Institute of Technology (IIT), Kharagpur, India. Currently, he is a professor in the Department of Electrical Engineering at IIT Delhi. He was also Dean, Undergraduate Studies at IIT Delhi. He was awarded INSA medal for young scientists in 1993. He is a fellow of Indian National Academy of Engineering and National Academy of Sciences, India. He is also a fellow of the International Association of Pattern Recognition. His research interests are in the areas of Computer Vision, Artificial Intelligence and Multimedia Systems. He has published more than 150 papers in international journals and conference proceedings. He has been on the programme committee of a number of international conferences like -ICCV, ACCV, ICPR, ICVGIP, PReMI etc.

Anupama Mallik did her B.Sc (1986) in Physics and Masters in Computer Applications (1989) from Delhi University, and Ph.D (2012) in Electrical Engineering from the Indian Institute of Technology (IIT), Delhi. Her Ph.D thesis dealt with ontology based exploration of multimedia contents. She is associated with Multimedia research group of the Electrical Engineering Department, IIT Delhi and has worked as a Research Scientist in projects sponsored by the Department of Science and Technology, Government of India. Her current research interests include Semantic web based applications, multimedia ontology, ontology applications in Internet of Things, and in cultural heritage preservation. She is a visiting faculty at the Indraprastha Institute of Information Technology, Delhi (IIIT-D). She is a member of the ACM.

Hiranmay Ghosh is a principal scientist with TCS Innovation Labs Delhi and leads its multimedia research track. He received his PhD from IIT, Delhi and his BTech and BSc degrees from the University of Calcutta. His current research interests include multimedia systems, knowledge representation and reasoning, semantic web, intelligent agents, cognitive models and robotics. He is an adjunct faculty member of the Electrical Engineering Department of IIT, Delhi and is associated with its multimedia research group. He is a member of the academic advisory committee of IIT, Ropar. He is a senior member of the Institute of Electrical and Electronics Engineers (IEEE), and members of ACM and IUPRAI.

Chapter 1

Introduction

1.1 The Multimedia Wave

The web is big, really big and getting bigger by the minute. The online population as of December 2013 was about 2.8 billion, which is about 39 percent of the world population. The estimated number of websites exceeds 650 million, with more than 50 million web-sites being added per year. Apart from the explosive growth in volume, there has been a significant change in the characteristics of web contents over the last decade. Much of the web is now proliferated by data contributed by the public at large. Further, the volume of multimedia contents on the web has far surpassed that of the textual data in the recent times. Over 300 million photos are uploaded on the Facebook every day and 4 billion hours of video have been watched on YouTube per month during the year 2012. [1] A few technology drivers that are responsible for this phenomenal change are

– *Web 2.0* has been responsible for proliferation of social networks. The users of the web are no longer passive consumers of data but actively contribute to the contents. A significant part of such data is in multimedia format, namely, photos, videos and music.

– *Mobile devices* such as smart-phones and tablets have built-in quality media-capture devices. These devices enable users to conveniently compose multimedia artifacts, although producing text artifacts still remains a legacy creative process.

– *Mobile access networks* enable users to access the web anytime anywhere. Users are no longer restricted to accessing the web from their working desks. This has led to higher participation in web-based activities. Trends indicate that there will be more online traffic from mobile devices than from traditional computers by the end of 2015.

Besides such crowd-sourced data, there exists a huge volume of professionally curated multimedia data collections in the form of television programs,

---

[1] Pingdom: Internet 2012 in numbers [royal.pingdom.com/2013/01/16/internet-2012-in-numbers]; Internet world stats: Usage and population statistics [www.internetworldstats.com/stats.htm].

space exploration archives and medical imagery, just to name a few. Moreover, Closed Circuit Television (CCTV) cameras installed at public utilities and industrial installations keep on generating a huge volume of video data every moment. The CCTV user group estimates that there are 1.5 million CCTV cameras installed in the United Kingdom alone, implying a camera for every 32 citizens in that country [76].

With this huge volume of multimedia data being available in different forms, their interpretation and semantic integration have become major research challenges. This book intends to present an insight into the challenges of large-scale semantic processing of multimedia data and the approaches to resolve them. This is aligned with the efforts to transit web technologies to a new level, known as as Web 3.0 or the *Semantic Web*, where the data on the web assumes machine-processable interpretation. Although there are several books on the subject, a major contribution of this book is to develop a unified ontology-based framework for interpreting multimedia data.

1.2   Semantic Multimedia Web: Bridging the Semantic Gap

There has been significant research in computer vision and audition technologies over the last few decades. But the *semantic gap* between the media features and the real-world entities that they represent has not yet been successfully bridged. In particular, variations in media forms and the differences in the methods used to interpret them in different contexts make formulation of a generic framework for interpreting web-scale multimedia data seemingly intractable.

In the wake of the 21st century, multimedia researchers have found the notion of *Semantic Web* quite interesting. Motivated by its success in the realm of textual document collections (often in multilingual format), researchers have turned to the Semantic Web technologies for contextual interpretation and integration of multimedia data. The core of Semantic Web technology lies in representation of the background knowledge about the domain of discourse in the form of *ontology* and to consider and interpret document properties in light of this background knowledge. The domain of discourse may, in general, include the application domain as well as the properties of the artifacts being dealt with. Several researchers have reported success in interpreting multimedia data with background knowledge about the domain as well as the media properties. Thus, Semantic Web technology is poised to migrate to the Semantic Multimedia Web [75].

1.3 Multimedia Web Ontology Language

While applying Semantic Web technologies in the realm of media data, a fundamental difference in the nature of media data and textual data has often been ignored. While text is an artificial artifact designed for communication of an event at the conceptual level, a media instance is a natural and perceptual recording of an event. The media events that are manifestations of conceptual events are marked by significant intraclass variations and uncertainties. The common knowledge representation techniques in Semantic Web technologies cannot represent such uncertain knowledge and deal with them. Further, a good part of media properties of a conceptual event is generally composed of the context that comprises related events in the domain context. The spatial and temporal composition of media events is also unique to the realm of multimedia data.

The differences between the representations of the conceptual and the perceptual spaces lead to the thought that there should be a fundamentally different way to represent knowledge about multimedia properties of concepts and to deal with them. In this book, we present a new paradigm for knowledge representation suitable for multimedia applications, embodied in a new ontology language "Multimedia Web Ontology Language" (or, MOWL in short). We have dealt with various aspects of encoding, learning and reasoning with this new knowledge representation paradigm in this book. In the wake of distributed big media data collections, as in the case of social networks and other large multimedia archives, autonomous distributed processing assumes significance. We present the new knowledge representation scheme as an enabler for such processing. Further, we have illustrated the use of the new ontology language over a wide cross-section of multimedia applications.

1.4 Organization of the Book

The rest of the book is organized as follows. Chapter 2 provides a brief overview of the evolution of the web, with an emphasis on Semantic Web technologies. Further, it deals with ontology representation schemes and their roles in creating domain models and semantic interpretation of information on the web.

In chapter 3, we characterize the semantics of multimedia instances, which is quite distinct from their textual counterparts. We distinguish the perceptual nature of media instances from the symbolic nature of text and establish the need for contextual interpretation of multimedia documents. We conclude the chapter with a review of the suitability of the current knowledge representation techniques for establishing multimedia semantics.

Chapter 4 provides an account of the recent research on ontology for multimedia data processing. In particular, we show how researchers have attempted to create a semantic description of media contents by combining multimedia content description in MPEG-7/MPEG-21 and ontology representation in OWL. This chapter discusses ontology applications in diverse domains such as ambient intelligence, biomedicine and phenomics. It concludes with a critical appraisal of these formulations and establishes why a traditional frame-based ontology representation scheme and crisp logical reasoning are not sufficient for multimedia applications.

We establish distinct requirements for an ontology representation scheme for multimedia applications in chapter 5. We move on to present a new ontology representation language, namely the Multimedia Web Ontology Language (MOWL), based on a framework for contextual reasoning with media properties for multimedia applications. The language provides mechanisms for encoding media properties of concepts in an ontology, for formal definition of spatio-temporal composition of complex events and for probabilistic reasoning with observed of media patterns. The chapter introduces the syntax and semantics of the new representation language as well as its reasoning framework.

Concepts encoded through multimedia ontology have to be linked with multimedia content for exploiting embedded semantics of multimedia data. In chapter 6, we discuss how content-based observable features can be used for modeling concepts associated with the data.

In chapter 7, we discuss the learning of a multimedia ontology and how this learning is different from the learning of text-based ontologies. We propose a multimedia ontology learning scheme that combines existing multimedia analysis techniques with knowledge that is drawn from concept-based metadata to learn a domain-specific multimedia ontology from a set of annotated examples. Examples from a classical domain illustrate how a more refined and robust version of a domain-specific multimedia ontology is built by learning from a dataset of conceptually annotated media documents.

In chapter 8, we discuss several multimedia applications pertaining to classification and retrieval, product recommendation and information integration. The common thread that binds these applications is that all of them use an ontology incorporating media properties of concepts. We illustrate with examples showing how such ontologies facilitate production of robust outputs with probabilistic reasoning in various contexts, even when partial inputs are available.

Chapter 9 presents an agent-based approach to building distributed multimedia applications with the proposed ontology representation scheme. We describe a mechanism to align and amalgamate independent and heterogeneous knowledge resources distributed over the Internet in different contexts using redundant media-based descriptions of complex media events derived from collection-neutral ontologies. We show how autonomy of the agents can be exploited to scale up a multimedia system to Internet scale using this new knowledge representation scheme. Further, we demonstrate the flexibility of the architecture with a few application examples.

Chapter 10 illustrates the application of multimedia ontology and its representation through the Multimedia Web Ontology Language (MOWL) in preserving tangible and intangible heritage through improved access of digitized heritage artifacts. Three diverse applications preserving cultural heritage with examples from Indian history and culture are discussed.

Chapter 11 summarizes the major contributions in the book and presents some open problems in the representations and applications of multimedia ontology. We conclude the book by arguing that MOWL has several desired properties to become the ontology language for the "Semantic Multimedia Web" that will be prevalent in the future.

Chapter 2

Ontology and the Semantic Web

2.1 Introduction

The explosive growth of web content has brought many information sources literally to the fingertips of Internet users, but not *sans ennui*; effective utilization of available information has become a significant challenge. The deluge of information has significantly increased the cognitive load on users in its processing. For example, several airlines may offer tickets at different prices for intercity travel, and several travel websites may compete with each other in offering "combo" offers with air-tickets, hotels and local taxi services. Exploring the websites to select an optimal offer takes significant time and effort on the part of a user.

The world would be a nicer place if some software on the net could understand the need of the user, interpret the information available on the different websites in that context and filter out unnecessary inputs. For example, if the user is on a flexible holiday schedule, the agent could discover the low-cost flights at convenient hours and hotels that suit his budget. Better still, the agent could infer what is best for the user and autonomously make the appropriate reservations and update the user's calendar. This remains an unfulfilled dream to date.

The root of the problem lies in the fact that machines do not have the capability of a human mind that intuitively interprets and relates bits and pieces of data. While the machines do process the contents on the websites and present them on the computer screen in a form suitable for human understanding, they cannot interpret their meaning by themselves. For example, the essential information about the book advertized in figure 2.1, namely the authors, the title, the year of publication and the price can be intuitively recognized by human beings but not by machines. Moreover, human beings can relate the title of the book with the subjects, that is, partitions of knowledge and the authors with their affiliations from alternate information sources. They can estimate the current relevance of the contents with the help of subject knowledge. Finally, they can make a purchase decision based on the information so gathered or deduced. Creating a travel itinerary or making purchase decision for a book by collating information from multiple sources and through logical inferencing in the context of specific user needs are complex and "intelligent"

FIGURE 2.1: A web page that can be intuitively interpreted by humans but not by machines

tasks that the machines are not yet capable of doing. The goal to make machines perform such intelligent tasks has led to research of a specialized set of tools, collectively known as the "Semantic Web", which is the topic of the current chapter of the book.

2.2   Evolution of the Semantic Web

The first generation of the web was designed with an objective of information sharing across the Internet. The contents of the web pages had been primarily designed for convenient human consumption. Static websites, meticulously designed by organizations and specific individuals, dominated the web. Setting up a website was a big deal and expressing oneself on the Internet was beyond the reach of the common man. Public at large had the role of being a passive consumer of information. Nevertheless, the web grew in size with time and discovering useful information by surfing became progressively difficult; search engines like Yahoo and Altavista then appeared on the web to facilitate information access. Web based e-mail services, like Hotmail, allowed the general users to exchange information with others for the first time over the web-based platform.

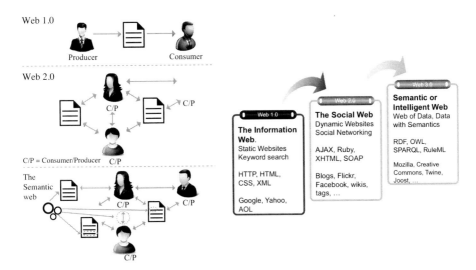

FIGURE 2.2: Development of the World Wide Web.

A major paradigm shift happened during early 2000s, when Web 2.0 or the interactive web enabled every individual to become an author. Users were empowered to contribute their thoughts in the form of blogs (web logs) on the web and publish photos or videos over a number of websites like Flickr and YouTube. Social networking sites like Facebook and Pinterest revolutionized social interaction norms. Twitter and several other micro-blogging sites helped in rapid dissemination of information. Web 2.0 led to a more serious form of information sharing and collaboration on the web too, as exemplified by the creation of knowledge assets like Wikipedia, which delegated traditional encyclopedias in print format to a page in history.

In more recent times, mobile networks have started offering large bandwidths to the customers and mobile handsets have started acquiring the features of a low-end computer. With a deep penetration of mobile handsets, especially in developing economies, the user base on the Internet has grown to a scale that is by far larger than ever before. Today, typical uses of the web include information retrieval (search engines, news, weather), social networking (blogs, Facebook, linkedIn, Twitter, Whatsapp), e-commerce (business-to-consumer, business-to-business), and so on. Figure 2.2 provides a crisp account of the development of the World Wide Web.

The contribution of the general public to the information base on the web made its volume grow in leaps and bounds. Data exist on the Internet in heterogeneous forms and in widely different formats. The large and heterogeneous information base has thrown open quite a few new challenges:

– Traditional search engines on the web no longer produce desired results. In order to provide access to a larger corpus of information, they generally achieve a high recall but a low precision value. Some of the search engines make use of sophisticated techniques in an attempt to improve search results but cannot keep up with the heterogeneity of information. To illustrate the point, a search for "Semantic Web" on a popular search engine yields 1.7 million web pages, many of which may not be useful to a typical user.

– Data buried deep in databases and other repositories cannot be generally discovered. Further, what is retrieved is sensitive to the way a query is formed and the exact representation of the information (rather than its semantics) on the web. The difference in the results obtained by the queries "Delhi Events" and "Events in Delhi" on Internet search engines illustrates the point.

– Human involvement is necessary to interpret and combine the information obtained from the Internet. The semantics of information obtained from the Internet cannot be understood by the machines and thus bits and pieces of information collected from multiple sources cannot be combined. As a result, information from multiple web pages cannot be collated to satisfy a complex information need, for example working out a travel itinerary, which is a common requirement for many users.

In order to make the vast information base on the web more suitable for human consumption, it is necessary for machines to be able to interpret, process, combine and summarize information on the web. The data on the web need to be viewed as a single repository rather than as fragments of webpages. Lee et al. articulated this view in their popular article in *Scientific American* way back in 2001 [15]. According to them, the "Semantic Web", which should make this possible, has yet to emerge, but when it does, the day-to-day mechanisms of trade, bureaucracy and our daily lives will be handled by machines talking to machines. The "intelligent agents" that the people have touted for ages will finally materialize. Since then, the vision has been extended to integrate real time data obtained from a plethora of sensors connected to the net with other forms of information and many researchers have talked about developing an "Intelligent Web", a "Wisdom web", a "Web of Things" or a "Wisdom Web of Things" [235].

Internet experts think that intelligent agents for the next generation of web, the Semantic Web, will be personal assistants that understand the users and will do all the hard work of interpreting the vast data on the Internet for his or her benefit. The services provided by such personal assistants will not be restricted to information gathering but will also to do more complex tasks like creating a travel itinerary for the user or making decisions on financial investments. While Web 2.0 uses the Internet to make connections between people, Web 3.0 or the Semantic Web will use the Internet to make connections between data and collate information.

In summary, the Semantic Web is all about linking data across applications. Human agents find it intuitive to extract and analyze information from heterogeneous and unstructured data sources. The world knowledge, knowledge of contexts and the experiences of human beings make this analysis possible. Machines or software agents, being deprived of such knowledge, cannot solve the problem of information extraction without explicit semantics. Semantic Web is focused on machines; web data is to be completely understood by machines, enabling them to search, aggregate and combine the information spread over the web without a human operator, in order to autonomously perform important high-level decision-making tasks. One way to make this possible is to complement the data available on the web with metadata and knowledgebases. Semantic Web is often characterized as "Web of Data"; figure 2.3 provides an illustration by Dan Zambonini.

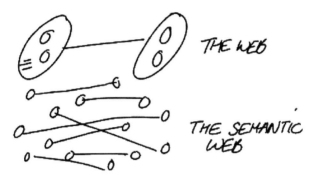

FIGURE 2.3: Semantic Web as a web of data (as illustrated by Dan Zambonini).

2.3 Semantic Web Technologies

In 2001, Tim Berners Lee outlined his vision of the Semantic Web and also presented a layered stack to represent the technologies that would implement and support the Semantic Web vision. By 2009, many new technologies had been added to this stack, and that is when Jim Hendler, one of the co-visionaries of the Semantic web presented his famous talk on Semantic Web technology stack, as a "layered cake" structure. The World Wide Consortium (W3C) has taken on the challenge of helping to build the "Web of Data" in addition to the classic "Web of Documents". They are responsible for defining the standards for the various technologies that are part of the Semantic Web stack. [1]

---

[1] See http://www.w3.org/standards/semanticweb/ for details.

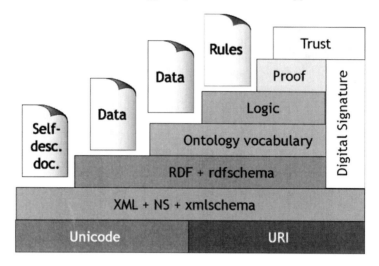

FIGURE 2.4: Semantic Web technology stack envisioned by Tim Berners Lee

The Semantic Web is based on the hypertext web. The first layer, on top of which its vision is built is composed of the web platform and its technologies, like HTTP (HyperText Transfer Protocol), and HTML (HyperText Markup Language) [208, 202]. The Semantic Web is a web of data, and its data have to have explicit semantics. In fact, every data element is a resource in Semantic Web, uniquely identified by a URI (Uniform Resource Identifier) or IRI (Internationalized Resource Identifier) [78, 77]. Semantics to the data are provided by adding metadata and by providing a structure. These are provided by the syntax of markup languages such as N3 (Notation 3), Turtle (Terse RDF Triple Language), RDFa, XML (Extensible Markup Language) [204, 13, 210, 203] and JSON (JavaScript Object Notation). [2] XML has its own set of technologies like NS (NameSpaces), DTD (Data Type Description), XPath, XQuery and XML Schema for providing access to the XML data.

Simple markup of data does not allow modeling of the world and the generation of new knowledge that is required in order to implement an "Intelligent Web". So the next layer in the stack is a knowledge representation structure supported by a simple knowledge representation language model of the RDF (Resource Description Framework) [209]. RDF can represent facts and assertions but is not expressive enough to model abstractions of the world, nor does it support logic for inferencing. Thus Semantic Web introduces the concept of ontologies (detailed in section 2.4 to model the world and infer new knowledge. This required advanced knowledge representation languages like RDFS (RDF Schema) [201] and OWL (Web Ontology Language) [211].

---

[2]http://www.w3schools.com/json/.

Ontology is a critical resource for Semantic Web applications. Reasoning with the knowledge helps to discover new facts that can be useful in certain application contexts. For example, knowledge of a user's budget and the price of flight tickets can be used to reason about the suitability of a flight while working out the itinerary of a user. Similarly, knowledge about the difficulty level of learning contents can be reasoned to ascertain their suitability for a student with a certain background. Ontologies are best suited to represent knowledge of a domain through modeling concepts, their properties and interrelations. Once the data are given a structure and knowledge has been captured through domain modeling in ontologies, access to both is provided by RDF query languages like SPARQL (pronounced "sparkle", an acronym for SPARQL Protocol and RDF Query Language) [206]. Semantic Web allows for knowledge to be encoded in rules and rule systems so that more complex notions like existential quantification, disjunction, logical conjunction and negation can be implemented. The RIF (Rule Interchange Format) [205] is a W3C recommendation. Its design is based on the observation that there are many "rules languages" in existence, and what is needed is to exchange rules between them.

A lot of work still needs to be done for the top three layers of the Semantic Web technology stack. For implementing security, a WebID protocol is being drafted. WebID [167] is an authentication protocol that uses the SSL/TLS layer for user identification engaging profile documents in security certificates. To make resource sharing safe in the Semantic Web, a CORS (Cross-Origin Resource Sharing) [207] standard is being developed by the W3C. Proof and trust in the Semantic Web are envisioned to be implemented through Named Graphs and Provenance XG. These are standards still under development with the W3C Semantic Web activity group.

Some well-known Applications of the semantic Web are the Social Web, DBPedia and LOD (Linked Open Data) cloud. The Social Web project [7] proposes to extend the interactions over social networking sites to a Semantic Web context by associating semantics to each interaction. It proposes a set of relationships that link together people over the Web. The Social Web intends to allow people to create networks of relationships across the entire web, while giving them the ability to control their own privacy and data.

Figure 2.5 shows the Linked Open Data cloud maintained by the Linking Open Data community project. It is based on metadata collected and curated by contributors to a data hub as well as on metadata extracted from a crawl of the Linked Data web conducted in April 2014. The goal of the W3C Linking Open Data community project is to extend the web with a data commons by publishing various open data sets as RDF on the web and by setting RDF links between data items from different data sources. DBpedia [216] is a crowdsourced community effort to extract structured information from Wikipedia and make this information available on the web. DBpedia allows you to ask sophisticated queries against Wikipedia and to link the different datasets on

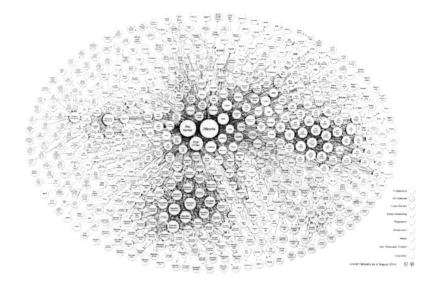

FIGURE 2.5: Linking Open Data cloud diagram 2014, by Max Schmacht-enberg, et al. http://lod-cloud.net/.

the web to Wikipedia data. Thus it reflects the vision of Semantic Web by linking structured data that have explicit semantics across the web.

2.4   What Is an Ontology?

Philosophers have been struggling to solve the riddle of the fundamental nature of the world since ancient times. The metaphysical debate on *what really exists* and *what it is like* has given rise to *ontology*, which literally means *study of beings*. The word is often used in a slightly different sense to denote "*a particular theory* about the nature of being or the kinds of things that have existence". [3] While scholars in different parts of the world have taken different views on the nature of existence, *ontology* as we understand today primarily evolves from the Greek philosophers like Pythagorus, Aristotle, Parmenides and Plato. The fundamental questions that these philosophers tried to answer are *what can be said to exist, what fundamental categories they may belong to* and *what are their properties.*

In modern times, research on ontology has been motivated by need for knowledge organization, representation and sharing in Library and Informa-tion sciences and in Artificial Intelligence (AI)-based systems. Gruber charac-

---

[3] from Merriam-Webster Dictionary (online).

terizes ontology as "an explicit specification of a conceptualization" [79]. In a computerized system, whatever can be represented can be said to *exist*. A *conceptualization* is an abstraction of the world that is created for some definite purpose. In the abstract view, certain aspects of the world that are considered unimportant in the context of the *purpose* are ignored. For example, while creating an ontology for a university, the curriculum is generally important, but the food preferences of the students may be ignored. The set of objects defined in a formal and explicit knowledge representation defines the boundary of the *domain of discourse*. The objects are organized into categories with formally defined properties.

Ontology uses a representation vocabulary, each term of which refers to an entity in the domain of discourse. There is an implicit association between an entity and its representation, which is generally a name or a symbol that can be intuitively understood by humans. The domain entity and its representation in the ontology need to be clearly distinguished. We shall see the importance of this distinction in later chapters of this book.

2.5   Formal Ontology Definition

An ontology represents the real world or a part of the world in terms of the entities that make up this world and the relations that exist between them. It captures the semantics of an entity through its attributes and its relationships with other entities. Thus an ontology should be able to describe the following elements:

- Classes or categories of entities in different domains of interest

- Relationships between these entities

- Properties or attributes that these entities may posses

Mark Ehrig in his book [52] has provided a formal definition of the ontology by a 4-tuple as

$$O := \{S, A, KB, Lex\}$$

consisting of the core ontology S, the L-axiom system A, the knowledge base KB, and the lexicon Lex. A core ontology or schema is the structure

$$S = (C, \leq_C, R, \sigma, \leq_R)$$

consisting of two disjointed sets C and R whose elements are called concept identifiers and relation identifiers (or concepts and relations for short); a partial order $\leq_C$ on C, called concept hierarchy or taxonomy; a function $\sigma : R \to C X C$, where $\sigma(r) = dom(r), ran(r)$ with $r \in R$, domain $dom(r)$, and

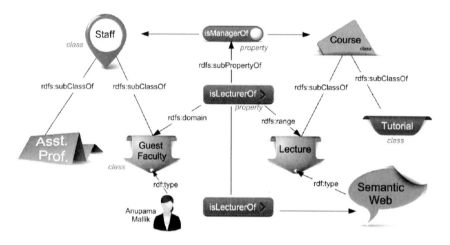

FIGURE 2.6: Graph view of an ontology snippet for an academic institution.

range $ran(r)$; and a partial order $\leq_R$ on R, called relation hierarchy, where $r_1 \leq_R r_2$ iff $dom(r_1) \leq_C dom(r_2)$ and $ran(r_1) \leq_C ran(r_2)$. If $c_1 \leq_C c_2$, for $c_1, c_2 \in C$, then c_1 is called a subconcept of c_2, and c_2 a superconcept of c_1.

Knowledge in the ontology can be expressed through axioms. Axioms are also used to apply logic for generation of new knowledge by reasoning with existing one. Thus axioms need to be expressed in a logical language, L. An L-axiom system for a core ontology is a pair $A = (AI, \alpha)$ where AI is a set whose elements are called axiom identifiers and $\alpha : AI \to L$ is a mapping. The elements of $A = \alpha(AI)$ are called axioms. Schema or core ontology S is considered to be part of the language L. The knowledge base KB extends the core ontology by adding instances or individuals that instantiate the concepts in the set C and adding relations between these individuals, which are seen as instances of relations in R. *Lex* simply represents the set of textual labels or signs for the entities in the ontology and the mapping between the entities and their labels.

Figure 2.6 shows a part of an ontology that models facts about an academic institution. The domain concepts refer to academic staff and courses with relations between them. As per the definition above, we have:

Set of Concepts $C := \{Staff, \; Course, \; Lecture, \; Tutorial, \dots\}$
Concept hierarchy $\leq_C := \{(Staff, \; GuestFaculty), (Course, \; Lecture) \dots\}$
Set of Relations $R := \{isManagerOf, \; isLecturerOf, \; \dots\}$
Function $\sigma := \{isManagerOf \to (Staff, \; Course),$
    $isLecturerOf \to (GuestFaculty, \; Lecture), \dots\}$
Relation Hierarchy $\leq_R := \{((isManagerOf, \; isLecturerOf), \dots\}$

This simple example shows that when knowledge is encoded in a formal specification like an ontology, then it has sufficient semantics to be machine processable. This entails that ontologies are the ideal fundamental basis for the Semantic Web.

2.6 Ontology Representation

Success of the Semantic Web depends upon data that has semantics and thus can be processed by machines. Semantics of data can be encoded in structured metadata with some formal specification. Early attempts towards semantic metadata led to development of markup languages like XML and XML Schema, which provided structure but lacked semantics. RDF provides the means to connect and share heterogeneous web data by treating everything as a resource and proposing a framework for its description and reuse in an ontology. RDF schema (RDFS) and then Web Ontology Language (OWL) added more semantics to an ontology and the capability to infer new knowledge. In this section, we discuss these technologies with an example demonstrating the evolution of ontology representation.

2.6.1 RDF and RDF Schema

W3C has been developing standards for the Semantic Web technologies. RDF is the W3C standard to express facts. Facts in RDF are expressed as statements that are triples with the following schema:

Resource (URI) + Property (URI) + Object/Value (URI/Literal)

where URI denotes Universal Resource Identifier. An RDF triple for "Anupama Mallik is lecturer of Semantic Web" can be expressed in N3 notation as

```
{
    http://www.example.com/academic/AnupamaMallik,
    http://www.example.com/academic/prop#isLecturerOf,
    http://www.example.com/academic/courses/SemanticWeb
}
```

For Semantic Web applications to use RDF data, a common sharable vocabulary of its terms needs to be defined. RDF schema defines a model for the creation of RDF statements, so is also called the *RDF Vocabulary Description language*. RDF Schema allows (a) definition of classes; (b) class instantiation in RDF via <rdf:type>; (c) definition of properties and restrictions; and (d) definition of hierarchies, that is subclasses and superclasses, subproperties and superproperties. An RDFS specification (expressed in Turtle) for the academic institution ontology is shown in figure 2.7. Using RDFS, we can

- declare classes like Staff, Course and Lecture

- state that GuestFaculty is a subclass of Staff

- declare isManagerOf as a property with domain Staff and range Course

- declare isLecturerOf as a property and state that it is a subproperty of isManagerOf

- state that AnupamaMallik is an instance of class GuestFaculty

Properties in RDFS can also have literals as values; for example, there can be a property hasPhoneNumber with Staff as its domain and a string as its range.

```
@prefix rdf: <http://www.w3.org/1999/02/22-rdf-syntax-ns#>
@prefix rdfs: <http://www.w3.org/2000/01/rdf-schema#>

:Course a rdfs:Class .
:Lecture a rdfs:Class;
        rdfs:subClassOf :Course.
:Tutorial a rdfs:Class ;
        rdfs:subClassOf :Course.
:Staff a rdfs:Class ;
        rdfs:subClassOf :Person .
:GuestFaculty a rdfs:Class ;
        rdfs:subClassOf :Staff .
:AsstistantProfessor a rdfs:Class ;
        rdfs:subClassOf :Staff .
:isLecturerOf a rdf:Property;
        rdfs:domain :Staff ;
        rdfs:range :Course .
:SemanticWeb a :Lecture .
:AnupamaMallik :GuestFaculty .
:AnupamaMallik :isLecturerOf :SemanticWeb .
...
```

FIGURE 2.7: RDFS specification of an ontology for academic institution.

2.6.2 Description Logics

Description logics (DL) denotes a group of knowledge representation formalisms that model the world in terms of concepts that belong to a domain and then uses this model to specify attributes and properties of objects and individuals that belong to the domain [8]. In view of the formal definition of an ontology, description logics can be used effectively to model an ontology. The family of logics is characterized by the use of constructs for building complex classes from simpler ones. They are different from first-order logic in that they provide tractable and decidable reasoning services.

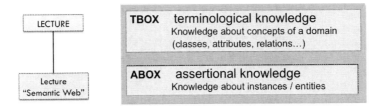

FIGURE 2.8: TBox and ABox in a DL specification of an ontology.

Ontological knowledge is represented in DL as TBOX or *Terminological knowledge*, which provides the vocabulary of the application domain, and ABOX or *Assertional knowledge*, which contains assertions about the entities and individuals that belong to the domain. DL supports the specification of:

– atomic concepts: `Staff`, `Lecture`

– atomic roles: `isManagerOf`, `isLecturerOf`

– constants: `AnupamaMallik`, `SemanticWeb`

– constructs like set negation(¬), union (⊔) and intersection (⊓) to define complex concepts; for example `GuestFaculty` are all those faculty who are guest staff at the institution `GuestFaculty = GuestStaff ⊓ Faculty`

– constructs to define complex roles; for example, when property `isManagerOf` has range `Lecture`, it defines a new property `isLecturerOf = ∃ isManagerOf.Lecture`

– Axioms: `GuestFaculty` is a subclass of `Staff`; that is `GuestFaculty ⊑ Staff`

– Assertions: Anupama Mallik is a lecturer of the course Semantic Web, as `isLecturerOf(AnupamaMallik SemanticWeb)`

It allows for reasoning for concept subsumption and concept instantiation. For example, if `HiranmayGhosh` $\xrightarrow{isLecturerOf}$ `DataEngineering`, then he must be an instance of class `Lecturer`. DL advocates crisp reasoning and has no support for uncertainty specification or probabilistic reasoning. This deficiency renders it unsuitable as a formalism for multimedia data and knowledge linked with it, requirements of which are discussed in later chapters of this book.

2.6.3 Web Ontology Language

Horrocks et al. [94] describes the evolution of the web ontology language OWL and the influence of Frames, Description Logics, RDF and several other formalisms on its philosophy and features. OWL borrows the basic fact-stating capability of RDF and structuring abilities of RDF Schema for classes and properties. Like RDFS, OWL can declare classes with a class hierarchy, but it extends RDFS capability by allowing combinations of classes through logical combinations (complements, union, intersection) of other classes. This is where it borrows from DL formalism. It also allows for classes to be declared as enumeration of individuals, which is not possible in RDFS. Extending RDFS, OWL can also declare characteristics of properties to say whether a property is transitive, reflexive, functional or the inverse of another.

Horrocks et al. also discusses how OWL borrows heavily from the model theory semantics of Description Logics, so that automated reasoning services can be performed to check the consistency of classes and ontology as well as to check the entailment relations [94]. Thus using OWL, we can additionally

- state that `Staff` and `Course` are disjoint classes

- state that `AnupamaMallik` and `HiranmayGhosh` are distinct individuals

- declare a class `AssistantProfessor` as an enumeration of three known assistant professors

- declare that a `GuestFaculty` can teach maximum two courses

- declare that an `AssistantProfessor` must teach at least one course

- declare that `isLecturerOf` and `isTaughtBy` are inverse properties

All these OWL declarations can also be expressed in DL. Figure 2.9 shows the academic institution ontology in OWL (using Turtle notation), with some of the above axioms encoded in it. Due to the various influences on OWL, there is sometimes a conflict between too much expressivity versus computational decidability. While RDF/XML syntax for OWL is too verbose and explicit, DL reasoning sometimes leads to undecidable logic. To balance these, W3C came up with three versions of OWL that can be selected to model an ontology depending on the application. These are

```
@prefix rdf: <http://www.w3.org/1999/02/22-rdf-syntax-ns#>
@prefix rdfs: <http://www.w3.org/2000/01/rdf-schema#>
@prefix owl: <http://www.w3.org/2002/07/owl#>


:Course a owl:Class .
:Lecture a owl:Class ; rdfs:subClassOf :Course.
:Tutorial a owl:Class ; rdfs:subClassOf :Course.
...
:Staff a owl:Class ; owl:disjointWith :Course.
:GuestFaculty a owl:Class ; rdfs:subClassOf :Staff ;
      rdf:type owl:Restriction ;
      owl:onProperty :isLecturerOf ;
      owl:minCardinality "2"^^xsd:nonNegativeInteger .
:AssistantProfessor a owl:Class ;
      rdfs:subClassOf :Staff ;
      owl:oneOf (:prof1 :prof2 :prof3) .
:isLecturerOf a owl:ObjectProperty;
      rdfs:subPropertyOf :isManagerOf ;
      owl:inverseOf :isTaughtBy .
_:x39 rdf:type owl:AllDifferent;
      owl:distinctMembers (:AnupamaMallik :HiranmayGhosh).
...
```

FIGURE 2.9: OWL specification of an ontology for academic institution.

– OWL DL: OWL expressed in DL having a friendly syntax and decidable inferencing.

– OWL lite: a subset of OWL DL with even simpler syntax and more tractable inference

– OWL full: a syntactic and semantic extension of RDFS with full expressivity and backward compatibility with RDF and RDF Schema.

Currently a new version of OWL called the OWL2 [211] is being developed by the W3C OWL Working Group.

OWL uses natural language constructs to represent ontology elements. In the formal definition of ontology in section 2.5, we saw how *Lex* represents the set of textual labels or signs for the entities in the ontology. This makes it convenient to apply OWL and other Semantic web technologies in the domain of textual information. Any attempt to utilize a conceptual model to interpret the perceptual records for multimedia data gets severely impaired by the semantic gap that exists between the perceptual media features and the conceptual world. Applications discussed in chapter 4 illustrate this fact.

2.7   Conclusions

We have presented the vision of the Semantic Web in this chapter. As requirements of Semantic Web are fully justified by an ontology, the latter is well suited to be the basis for Semantic Web implementation. We have provided a brief summary of the Semantic Web technologies and their evolution. We have defined an ontology formally and shown a sample ontology specified in various ontology representation schemes. We argue that though OWL provides a sophisticated ontology representation scheme with a crisp reasoning framework, it is more suited to encode ontologies with textual data. Multimedia semantics have more complex requirements, which are discussed in detail in chapter 3.

Chapter 3

Characterizing Multimedia Semantics

3.1 Introduction

Today, the Internet hosts many image and video collections related to current news, sports, medicine, digital heritage, and other topics in the public domain. This vast collection is a huge source of knowledge and can be used in various ways, from searching for some music for entertainment to learning the characteristics of a specific type of tumor for research purposes. Discovering, correlating and characterizing the relevant media instances in a specific context is, however, a daunting task. Generally, a semantic description of the media contents, or what they *signify*, is more important to a user than its physical format and visual attributes. The challenge of imprecise association between the media attributes and their connotations is compounded by the different contexts of media usage and the variety of media forms.

There has been strong research interest in extraction of semantics of audiovisual documents for more than a couple of decades now. Early works on media semantics had been confined to small collections of images, music and video in narrow domains. These efforts do not scale up to distributed and heterogeneous collections on the Internet. In the previous chapter, we saw that ontology, or formal domain representation of domain knowledge, is an important tool for contextual semantic interpretation of documents in an open Internet environment. The explicit and collection-neutral encoding of domain knowledge is the key to building web-scale semantic applications. With the proliferation of media content on the web, several research groups have attempted ontology-based interpretation of multimedia document contents. Despite such efforts, multimedia semantics still remains an elusive concept. The *semantic gap* between media features and concepts they represent is yet to be successfully bridged.

An appreciation for the characteristics of multimedia semantics is essential for understanding these approaches and their shortcomings. The essential difference of multimedia instances from textual documents is that the former carries information in perceptual form, while the latter encodes it in a symbolic form. This makes multimedia semantics a much more intricate issue than its textual counterpart. In the rest of the chapter, we present the perceptual nature of media contents and the different facets of their interpretations.

3.2   A First Look at Multimedia Semantics

Characterizing the semantic content of a media instance has many flavors. Sometimes, it is the objects or the personalities depicted therein that are important. On some other occasion, the depicted objects melt away in the scene and the overall impression that it conveys becomes important. In this section, we try to analyze the many faces of multimedia semantics before formulating methods to interpret them.

3.2.1   What a Media Instance Denotes

The lowest level of media semantics is possibly captured through the items in pictorial depictions. For example, if we look at the images on the different rows in figure 3.1, our first reaction can be to state that they denote a monument, a specific person or a class of objects, namely the the Taj Mahal, Narendra Modi (present prime minister of India) and flowers. The media instances have significant variations in media patterns that represent these "concepts" because of several environmental factors. The human cognition process can robustly recognize an object despite such variations. Further, the objects of interest do not cover the entire image area. Although there are significant visual patterns that do not constitute those objects, human mind has no difficulty in filtering the background. Similarly, a speech or a song generally contains some words representing a few key concepts, which a human mind can filter from the rest.

FIGURE 3.1: Identification of objects.

3.2.2 Interaction between the Objects and the Role of Context

The next level of semantics in media artifacts arises out of the interaction of objects in a multimedia document. A few examples of such interactions are shown in figure 3.2. The images on the top row show interactions of human beings with each other and with some inanimate objects, for example, ball, goalpost, and so forth. The images can be interpreted to represent some episodes of football matches based on observation of such interactions. Similarly, the interactions of human beings with different musical instruments signify musical performances in the images on the lower row. We observe that the knowledge about the context plays an important role in determining the semantics of media content. For example, the presence of the symbol of a political party and his presence in an official ceremony in two of the images in figure 3.1 help in identifying Narendra Modi to someone who may not be very familiar with his face. The contextual association demands some background knowledge about Mr. Modi. Similarly, knowledge about a football match makes us believe in the possibility of a goal being scored, when we see a ball in the vicinity of the goalpost and hear the noise of excitement from the audience.

FIGURE 3.2: Interactions between objects (images are borrowed from Wikimedia Commons).

3.2.3 The Connotation of Media Forms

The semantics of media instances generally go beyond the objects and their interactions. The real semantics of multimedia lies in the impressions that it evokes in a human mind. In many instances, a media instance is specifically curated by its creator to communicate a specific viewpoint to the audience. For example, a photographer uses the camera controls to capture and present his subjective and aesthetic view of the world. A performer of classical dance enacts her subjective interpretation of the accompanying music. She tells a story through her body postures, hand gestures, facial expressions and foot movements. TV newscasts presents a viewpoint of an event by selective editing of the audiovisuals and complementing them with the anchorperson's interpretation.

Going further, media instances can also be used to create an impact on human emotions. Such techniques are extensively used in various art forms and in the advertising industry. For example, photographs of nature that may evoke a sense of relaxation or peace are generally used by advertisers of holiday resorts. Certain sound patterns that create a sense of fear and suspense are extensively used in horror movies. Thus, the semantics of media takes a deeper meaning in the curated media forms. It is no longer what is being depicted in the media instances, but what they connote. Connotations of media forms have a large role to play in all forms of arts and crafts as well as in media and advertising industry.

3.2.4 The Semantics Lie in the Mind of the Beholder

FIGURE 3.3: Multiplicity of interpretations (image borrowed from Wikimedia Commons; http://en.wikipedia.org/wiki/File:Van-willem-vincent-gogh-die-kartoffelesser-03850.jpg).

The semantics of a media form that needs to be exploited is often subjective and depends on the specific context of use. A journalist may be looking for a photograph of a specific person or a video clipping for a specific event. In such case, the semantics of the media form is established by the depiction of that particular person or the specific event in the media instances. On the other hand, an advertiser of a brand of edible oil may want to use an image that will convey the notion of a healthy family. In such case, the actual contents of the image – that is the identity of the human beings or their interactions being depicted – are unimportant. Further, observers may interpret the same media instance differently depending on the context. For example, the painting in figure 3.3 may be interpreted either as a depiction of the lifestyle of European farmers in the nineteenth century from a sociological perspective or as an example of Renaissance painting from the perspective of an art lover.

3.2.5 Multimedia as a Communication Channel

Taking a cue from the previous discussions, we can view multimedia artifacts as a communication channel through with an author communicates with a consumer, as shown in figure 3.4. An author creates a multimedia instance with some intent. A consumer interprets it with his or her perspective. Thus, the semantics of a media instance is guided not only by its contents but also by the intent of its creator and the interpretation by the observer. Thus, it is not possible to represent the semantics of multimedia artifacts by describing its contents alone, for example, as a spatial and temporal arrangement of objects that it may contain. The interpretation of a media instance by an observer is guided by the observer's world model. A shared world model between the creator and the observer enables effective communication. In addition to the media contents, this shared world model is also an essential ingredient of multimedia semantics.

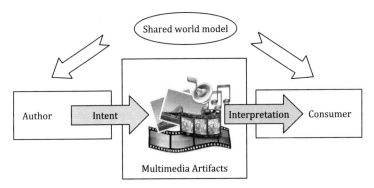

FIGURE 3.4: Multimedia as a communication channel.

3.3 Percepts, Concepts, Knowledge and Expressions

3.3.1 Percepts and Concepts

A human being experiences the real world through sensory organs. The sensory input is processed by the brain to create a mental model of the environment. This mental model is called the *percept*. The process of perception involves identification, organization, correlation and interpretation of sensory inputs to create a representation of the world that a person is exposed to. The sensory inputs may either lead to a passive reception (for example, just gazing around) or some instantaneous reaction (for exampe, to a loud bang). Perception is a step forward, where the sensory inputs are analyzed and refined to model the world – for example, shapes to distinguish real-world objects. This is also associated with learning. For example, a person may see several instances of monuments of different descriptions. Each instance of viewing results in the reception of visual signals and creation of a mental model. An analysis of such mental models results in discovery of some common visual patterns – for example, existence of round domes and circular arches across some of these monuments, albeit their differences as shown in figure 3.5. These common patterns represent a *concept*, that is, a class of real-world entities. A mental analysis of similarity and dissimilarities of the observed patterns leads to concept hierarchies, for example, monuments, tombs, forts and palaces. Spatial or temporal co-occurrence of patterns leads to discovery of contextual relations between the concepts and realization of complex concepts. For example, discovery of the temporal co-existence of certain body movement patterns and audio-patterns results in the contextual association of dance with music. Achievement of some mental states on experiencing some sensory inputs leads to realization of *abstract* concepts, like joy and fear.

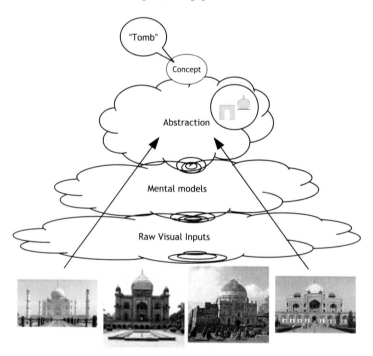

FIGURE 3.5: Abstraction of mental models (images of the monuments are borrowed from Wikimedia Commons).

3.3.2 The Perceptual Knowledge

Concepts are abstractions of observed media patterns. A corollary that follows is that each of the concepts is represented as some sensory patterns in a human mind. Concepts are related with each other by virtue of the similarities and dissimilarities of the sensory patterns that they represent and their spatial or temporal co-occurrence. A person's knowledge about the world is a collection of the concepts, their association to sensory patterns and their interrelations, as experienced by that person. This knowledge is essentially perceptual in nature. New sensory inputs are constantly compared with the perceptual world knowledge of a person to make inferences about the real-world entities around him or her and also to update the knowledge. The knowledge about the world so acquired leads to *attention*, which helps human minds to selectively process some sensory signals that are important, for example, to speech in a noisy background or to a shiny Jaguar (a car) amidst the clutter of the traffic on a busy road, and ignore the others in specific contexts. In this context, it is worth noting that the percepts acquired by a human being that arise out of experience are strictly private and are not shared with anybody else. Thus, a person dwells in *his* or *her* own world and interprets the world from *his* or *her* viewpoint. The relation between the percepts of a person, often referred to as *Maya* or illusions, and the reality that *really* exists is a matter of fierce philosophical debate. We shall refrain from that debate in this book. We conclude this section with an observation that, strictly speaking, there is no common interpretation of the sensory inputs across human beings.

3.3.3 Communication and Expressions

With the advent of civilization, humans have evolved various means for communicating percepts. Various symbols have been designed for effective and concise communication of mental models. Birds and animals can communicate limited percepts – imminent danger or source of food – to each other by emitting some sound patterns or through some body gestures. It is believed that emission of such sound patterns evolved into spoken languages for humans, though scientists differ in their opinion about the mechanism of such evolution. Each concept is represented by a specific sound pattern in speech. The act of utterance of a sound pattern denotes an intent to communicate a set of related concepts to an intended audience. Such utterances and their receptions lead to a shared knowledge structure across the participants that helps in effective communication. Some of such associations get genetically encoded and provide a bootstrapping mechanism for such sharing.

Communication through sound is transient and does not exist once it has been completed. For recording of experience, humans resorted to drawing. Early cave paintings show pictorial symbols being used to express human experience. This is possibly the earliest evidence of written communication. Each picture in a cave painting is a symbolic representation of a real-world object class, such as animals, humans and weapons. A spatial arrangement of such pictures represents real-world events like hunting and war. The auditory and pictorial expressions matured over the ages to various audiovisual art forms, such as music, painting, drama and cinematography, a major source of multimedia artifacts in the modern age. In another stream, they matured into speech and written text and natural languages, refined with some artificial ingredients such as the grammar, as we know them today. A word connotes a certain concept, which in turn represents some sensory patterns. For example, the word "tomb" represented by the sequence of letters t, o, m and b, refers to a concept that signifies certain spatial arrangements of a few shape primitives as shown in figure 3.5. A linear sequence of words in a modern natural language is a symbolic representation of real-world concepts and their interactions.

The technology to integrate multiple media forms in a common representation framework has led to expressions in multimedia format, where a combination of text, graphics, video, speech and animations are used to convey a set of concepts and their interactions. The documents use the different media forms to make the best use of perceptual and symbolic representations to articulate the author's viewpoint. The media forms are generally synchronized to maintain a conceptual flow over the communication medium. In many instances, the multimedia documents are created in interactive formats to engage the viewers with the contents, leading to a more effective channel for communication. Figure 3.6 depicts various modes of expression and communication in perceptual and symbolic forms.

3.3.4   Symbolism and Interpretation

We find that human communication primarily takes place either through a natural language (written text or speech) or through media artifacts, such as images and video. Since the percepts are strictly private to a human mind, all forms of communication need to be symbolic. In speech, specific sequences of audio patterns refer to some concepts. In written text, specific visual patterns represent letters of an alphabet, a sequence of which represents one or more concepts. The non-speech audio and non-text visual patterns in curated multimedia artifacts are designed to convey some percepts.

Looking deeper into the symbolism used in communication, we see that some symbols are formally defined, as in the case of mathematical, logical expressions and the programming languages. Their interpretation follows some well-defined rules that are commonly understood and accepted in the community. Thus, they become strictly formal modes of communication. Formal symbolism in the form of graphics are used in road signs, musical notes and the emoticons used in social media. The common use of natural language, as in our everyday conversation, is also a formal representation of concepts, though the formalism as defined by the grammar of the language is weaker than those of the programming languages. A statement such as "The Earth is round" conveys a proposition that is generally well understood without

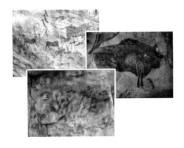

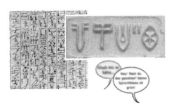

(a) Cave paintings

(b) Towards symbolic representation: icon and speech

very large (usually about 10,000 different words) vocabularies. I
of vowel and consonant speech sound units. These vocabularie
creating the existence of many thousands of different types of n
communicate in two or more of them,[3] hence being polyglots.
humans with the ability to sing

A gestural form of human communication exists for the deaf in t
written language, often one that differs in its vocabulary, syntax
Speech is addition to its use in communication, it is suggested t
processes to enhance and organize cognition in the form of an

Speech is researched in terms of the speech production and sp
concern speech repetition, the ability to map heard spoken wor
vocabulary expansion in children and speech errors. Several ac
pathology, linguistics, cognitive science, communication studies
the human brain in its different areas such as the Broca's area i

It is controversial how far human speech is unique in that othe
have comparably large vocabularies, research upon the nonverb
the possibility that they might have these capabilities. The origi

Patna railway police seems to have committed gr
filed after the first of the serial blasts in Patna. Go
Police (Patna) station house officer stated in the F
"could not produce valid documents for keeping
it".

(c) Text

(d) Multimedia: Symbolic and perceptual

FIGURE 3.6: Modes of communication in perceptual and symbolic forms (images are borrowed from Wikimedia Commons).

any ambiguity. Ambiguities arising out of use of pronouns and words with multiple meanings, such as "Jaguar", are generally resolved with contextual interpretation.

Language has, however, been used in literature in different ways. An idiom such as "Leopards cannot change their spots" has a meaning that has nothing to do with the leopards (an animal). In Thomas Gray's *An Elegy Written in a Country Churchyard*, "Full many a gem of purest ray serene ... " refers to the people with intellectual potential who were born and died in the anonymity of a village. Such symbolism in literature is imprecise and its interpretation is extremely subjective.

Coming to the world of multimedia, a matter-of-fact depiction – for example, a video recorded by a surveillance camera – conveys almost the same sensory inputs as if the viewer were present at the spot. Similarly, a painting belonging to the naturalist school depicts realistic objects in their natural settings and attempts to generate the same sensory experience in the viewer, as if she were actually subjected to the scene. Unlike natural or formal languages, there is no "grammar" for the naturalist depiction in media forms. Like in literature, symbolism is also prevalent in the media domain. In symbolic art forms, the objects depicted often represent things other than themselves. In more abstract form, an artist relies on low-level media features such as colors and textures, rather than objects, to convey an emotion. For both naturalist and symbolic depictions, the interpretation of media forms depends on the perceptual world-model in the mind of the viewer.

So, how do we interpret a symbolic communication without formal semantics? The communication channel, or the multimedia artifacts, present a set of sensory inputs to the receptor. The human mind, when confronted with new sensory inputs, explores acquired perceptual world knowledge to find the best possible explanation for the observed sensory inputs. The semantics lie in the holistic interpretation of the received signals in their entirety and not a sum of their piecewise interpretations. Analysis of individual signals in the context of other received signals makes the interpretation robust against incompleteness and uncertainties in the scene. The reasoning process in human cognition is characterized as plausible reasoning [108], also known as *abduction*, that is, inferencing by evidence.

3.4   Representing Multimedia Semantics

The meaning of a natural-language statement is established by the meanings of words it contains and their sequence. As a "word" is an atomic semantic entity in a natural-language statement, an "object" can be treated as an atomic semantic entity in multimedia instances. Following the approach in natural-language processing, the semantics of a multimedia instance could be represented as the objects that it contains and their spatial and temporal arrangement. For example, we could define

$$dome \land arch \land above(dome, arch) \rightarrow tomb$$

where *dome* and *arch* represent visual patterns, *above* represents a spatial relation and *tomb* represents a concept signified by a spatial representation of the visual patterns.

This method of representation, even in the simple case of monuments having well-defined structures, has quite a few shortcomings. If we look at the third image in figure 3.5, the arches are not distinctly visible and there is every likelihood of an object recognition algorithm failing to detect them. Although the human visual system can readily recognize the similarity of the monument with the others, the inferencing mechanism – where the whole is recognized as the sum of the parts – to detect the concept is likely to fail in this case. Such failures can be caused by many other reasons, such as (partial) occlusion, difference in viewpoints, illumination differences, and so on. Figure 3.7 illustrates some of such examples. It can be argued that all such variations can be encoded in a set of rules, but the combinatorial explosion of premise variables, even for a simple example like a monument, makes encoding such rules practically impossible. More than the numbers, it may not be possible to identify or predict all such combinations and encode them in advance. A more fundamental problem with such representation is that it cannot deal with symbolism, that is, connotations of multimedia objects. Although it may be possible to identify the Taj Mahal through its structural description, there is no way to associate it to the *romanticism* that it symbolizes. Note that this problem is not particular to multimedia but is common to natural-language text also. There is no way to understand an idiom or a poem by a structural analysis of the text.

FIGURE 3.7: Occlusion, viewpoint and lighting differences for the same monument (images are borrowed from Wikimedia Commons).

The representation of semantics through structural composition of constituent elements, whether in text or in multimedia, suffers from the defect that it accounts for the contents but not the world knowledge of the observer. It may be argued that this world knowledge is used in deciphering the semantics of the constituent components – namely, the words in natural-language text or the objects in a multimedia artifact. This argument misses the point that the interpretation of the whole is not the sum of interpretations of the components.

Establishing semantics of multimedia data (including symbolic text) is all about cognition. It needs to emulate how the human mind interprets sensory inputs as a whole, in the context of each other. Thus, the semantics of multimedia artifacts lies in the evidential value of the embedded media patterns towards the concepts. In order to determine the semantics of a multimedia artifact, we need to compare the media patterns with the perceptual models of the concepts and try to arrive at the best match. We shall elaborate on the perceptual modeling for multimedia data interpretation in a later section.

3.5   Semantic Web Technologies and Multimedia

In his seminal paper [15], Lee et al. proposed a vision where all data and services on the web can be interpreted by machines for convenience of human consumption. This led to the development of a set of technologies collectively known as the Semantic Web technologies to enable such machine interpretation. Ontology, which is a formal representation of the knowledge about a domain, is a critical component in the semantic interpretation of documents. An ontology encodes hierarchical concept structures and the properties of the concepts that characterize a domain. The ability of formal reasoning with the domain representation enables semantic characterization and correlation of different documents in context of that domain. This was described in detail in chapter 2 of this book.

The Semantic Web vision encompassed *all* forms of data. The logical domain model described above is a symbolic representation of the domain and is, in principle, separated from any specific media form. However, the Semantic Web technologies developed in early 2000s were primarily engineered towards understanding textual data that dominated the web during that time. Since natural-language text has been closest to the conceptual representation compared to any other available form of communication, it has been a natural choice for representing a domain knowledge. Natural-language constructs have well-established meanings (with the exception of idiomatic use), which makes them suitable to represent the elements in an ontology – namely, the concepts and the relations. Readily available hardware and software technologies – for example, the computer keyboard, displays, ASCII coding scheme, text editors, the dictionaries and the *Wordnet* – added to the convenience of its use. Formalism has been established by complementing the natural-language constructs with logical structures and rules. Thus, the Semantic Web technologies and the supporting tools as we know them today are mostly text oriented.

In contrast to text documents, multimedia documents are sensory representations of natural events. There is a gap between the concepts, expressed through natural language in an ontology, and the media patterns in the media instances. A concept is an abstraction of percepts formed out of the sensory inputs. Thus, in principle, it is possible to encode some mapping between the concepts and the media patterns. However, such association is extremely noisy, a phenomenon that is referred to as the *semantic gap* in the literature. An attempt to interpret sensory representations using the logical framework of an ontology gets severely impaired because of the noisy association.

We contend that the existing concept modeling techniques are not suitable for interpretation of multimedia document contents. In order to cater to the multimedia document collections on the web, Semantic Web technologies need to be extended to create a *perceptual* domain model that contextually associates the conceptual world with the perceptual world and bridges the semantic gap. We distinguish the perceptual modeling from the existing conceptual modeling technique in the following section and arrive at the requirements for a knowledge representation scheme for semantic multimedia applications. We shall elaborate on this contention in the later chapters of this book.

3.6 Conclusions

We have characterized the nature of multimedia semantics in this chapter. We have seen that concepts can manifest in many different forms in multimedia documents and that contextual knowledge is essential in multimedia content interpretation. We have shown that the context depends on several factors, such as the experience of the consumer, interaction between the objects depicted, domain of discourse, the relationship between the media segments, and most importantly the personal viewpoint of the consumer. We have analyzed symbolism in multimedia, when it serves as a channel of communication, and the information leak that may happen across the channel. We have briefly commented on the challenges in the application of Semantic Web technologies for multimedia data interpretation. We review a significant body of related work in chapter 4, which illustrates the issues. We expand on the incompatibility between present ontology representation schemes and a possible approach to solve the problem later in this book.

Chapter 4

Ontology Representations for Multimedia

4.1 Introduction

Ontology is the philosophical study of being. It is concerned with everything that is real. Fundamentally, ontology deals with existence, identity, part, object, property, relation, fact and the world, building the framework for specification of conceptualization for a domain. Ontology, in this sense, provides the foundational principle for interpretation and conceptualization of multimedia content, which in a concrete form records everything perceivable through human senses and hence is linked to the reality of existence and the world. The representation for a concept is to be constructed through a hierarchy of descriptors and linguistic references built upon raw multimedia data. Standardization of the descriptor set as attempted in MPEG-7 provides the basic terminological layer for describing multimedia content. Ontological layers have been constructed on top of these standardized descriptors. In this chapter, we examine these representations. We also identify basic conceptual problems with these approaches.

4.2 An Overview of MPEG-7 and MPEG-21

The objective of MPEG-7 [134] was to define a set of standardized tools to support the generation of human understandable audiovisual descriptors that can be used for fast and efficient retrieval from digital archives (pull applications) as well as filtering of streamed audiovisual broadcasts on the Internet (push applications). The main elements of the MPEG-7 standard are:

- Descriptors (D) are representations of features. Descriptors are provided for defining the syntax and the semantics of each feature representation

- Description Schemes (DS) specify generic and multimedia description tools. These tools allow creation of the structure of the descriptions of individual media elements, collections and user preferences, with the provision for including descriptors and other description schemes

- Description Definition Language (DDL) is an extension of the W3Cs XML Schema language with the addition of MPEG-7 specific extensions. It provides syntactic specification of a language for creation of new Descriptors, new Description Schemes, and extension and modification of standardized Description Schemes. XML Schema based DDL provides little support for expressing semantic knowledge

- System tools are defined for supporting multiplexing and synchronization of descriptions with content. They also specify transmission mechanisms and coded representations (both textual and binary formats) for efficient storage and transmission, management and protection of intellectual property in MPEG-7 descriptions.

The descriptor set of MPEG-7 includes visual descriptors, which describe basic audio-visual content of media based on visual information. MPEG-7 provides descriptors for describing image and video content. These descriptors can represent, for example, the shape of objects, object size, texture, color, movement of objects, and camera motion. These descriptors need to be extracted from the image or video data. These descriptors are linked or embedded with basic audiovisual content. In the following, we examine standardized representation scheme of some of the descriptors:

- Color is one of the most widely used visual features in image and video retrieval. Color features are independent of image size and orientation. To allow interoperability between various color descriptors, normative color spaces are constrained mostly to the well-known Hue-Saturation-Value (HSV). Some of the Color descriptors defined are:

  - Scalable Color Descriptor (SCD): This feature descriptor captures basic color distribution in an image. The MPEG-7 generic SCD is a color histogram encoded by a Haar transform. It uses the HSV colors space uniformly quantized to 255 bins.

  - Dominant Color Descriptor (DCD): In contrast to the color histogram approach, this descriptor provides a much more compact representation. Colors in a given region are clustered into a small number of representative colors. The descriptor consists of the representative colors and their percentages in a region.

  - Color Layout Descriptor (CLD): This descriptor describes spatial distribution of color in an arbitrarily-shaped region. Color distribution in each region can be described using the Dominant Color Descriptor defined above.

– Group-of-Frames/Group-of-Pictures (GoF/GoP) Color Descriptor:
The GoF/GoP color descriptor defines a representation scheme for
representing color features of a collection of similar frames of a
video by means of the SCD.

– Visual Texture Descriptors: Texture refers to the visual patterns due
to distribution of multiple colors and gray-level patterns. Textural patterns are visible in images of virtually any surface, including clouds,
trees, bricks, hair, and fabric. MPEG-7 has defined a number of texture descriptors that can be employed for a variety of applications. For
example,

– Homogenous Texture Descriptor (HTD) describes directionality,
coarseness, and regularity of patterns in images. It is most suitable
for a quantitative characterization of texture that has homogenous
properties. The descriptor is based on a filter-bank approach employing scale and orientation sensitive filters.

– On the other hand, MPEG-7 defines Edge Histogram Descriptor
(EHD) to provide descriptions for non-homogenous texture images. This descriptor captures spatial distribution of edges. The
extraction of this descriptor involves division of the image into 16
non-overlapping blocks of equal size. Edge information is then cal-
culated for each block in five edge categories: vertical, horizontal,
45 degree, 135 degree, and non-directional edge. It is expressed as
a 5-bin histogram, one for each image block.

– Visual Shape Descriptors: MPEG-7 provides for shape-based descriptors
for the image. The shape descriptors are expected to be invariant to scaling, rotation, and translation. Shape information can be 2-D or 3-D in
nature, depending on the application. In general, 2-D shape descriptors
can be divided into two categories, contour- based and region-based.

– Motion Descriptors for Video: All MPEG-7 descriptors described above
for color, texture, and shape of objects can be readily employed to index
images in video sequences. Description of motion features in video sequences provides more powerful indication regarding its content. MPEG-
7 has defined descriptors, which describe characteristics due to the camera motion and object motion. Given a group of frames, its overall activity level or pace of motion or action in a scene is captured by Motion
Activity Descriptor defined in MPEG-7 standard. This descriptor describes whether a scene is likely to be perceived by a viewer as being
slow, fast paced, or action paced. An example of high activity video
sequence is a scene in a soccer match.

MPEG-7 has the provision of defining segments in the multimedia content
and assigned descriptor-based abstract description schemes to the segments.
Some example description schemes are:

- Still Region Description Scheme

- Video Segment Description Scheme

- Audio Segment Description Scheme

- Audiovisual Segment Description Scheme

- Collection Description Scheme.

Some of these description schemes have a taxonomic relation between them. For example, Still Region, Video Segment and Audio segment Description schemes extend the Segment Description Scheme. We show an example of a video frame with different segments (SRi) and associated descriptors in figure 4.1. A snapshot of a video with two temporal segments is shown in figure 4.2. Each region contains several regions (still and moving). The Segment Relationship Graph description tool shown in figure 4.3 is used to describe the composition and positional relationship of regions within the same segment and the relationship of the same region from one segment to another.

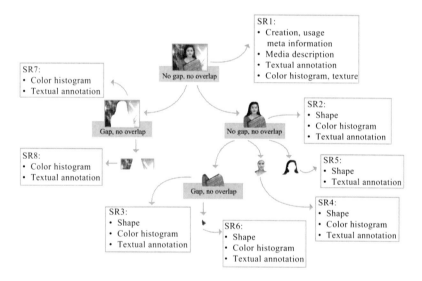

FIGURE 4.1: MPEG-7 segments and their descriptors for a video frame.

MPEG-21, the open framework of multimedia applications, is an attempt to standardize and define the glue technology needed to support users to create, manipulate, trade and consume multimedia content in an efficient way. The media information is abstracted and encapsulated in units called "Digital Items". Digital items are generated and/or consumed by "Users" interacting between themselves. Digital items are structured containers with a standard

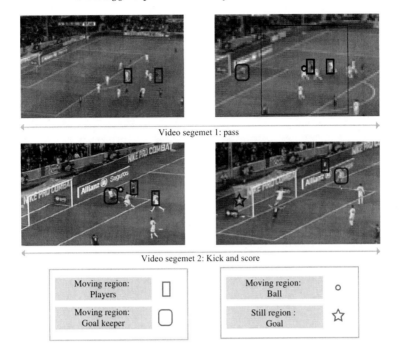

Video segemet 1: pass

Video segemet 2: Kick and score

| Moving region:
Players | ⬜ |
| Moving region:
Goal keeper | ⬭ |

| Moving region:
Ball | ○ |
| Still region :
Goal | ☆ |

FIGURE 4.2: Example of video segments and regions.

representation including resources and metadata, for example, a movie accompanied with the soundtrack, statements by an opinion maker, ratings of an agency, and so forth . Among other things, it also includes identifiers, licenses written in a "Rights Expression Language" (REL) and intellectual property management and protection information. The latest component of the MPEG-21 is the "Media Value Chain Ontology" (MVCO). The MVCO is the ontology for formalizing the representation of the media value chain. The media value chain is the process by which a work is conceived, represented, distributed or broadcasted and consumed. The MVCO focuses only on those actions that are relevant for management of intellectual property. There have been other attempts to define vocabularies to standardize description of the content. Some examples are:

- TV-Anytime (http://www.tv-anytime.org/) standardizes 954 terms for broadcast TV content.

- Thesaurus of Graphical Material (TGM-I) defines index terms for cataloging graphical (image and video) content.

- SMPTE Metadata Registry from the Society of Motion Picture and Television Engineers is a metadata dictionary. It is a registry of metadata element descriptions.

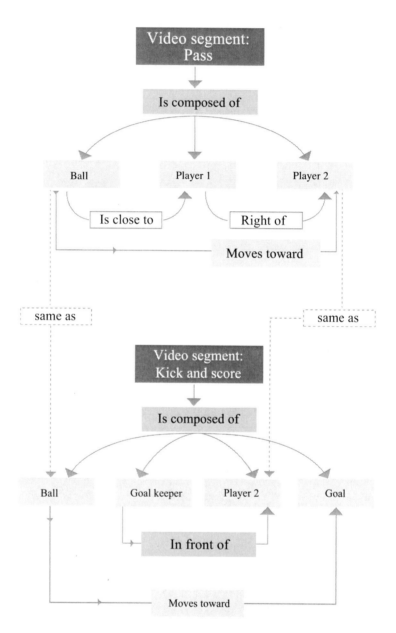

FIGURE 4.3: Relationship graph for video segments.

4.3 MPEG-7 Ontologies

In MPEG-7 multimedia content is classified into five types: image, video, audio, audiovisual and multimedia. MPEG-7 also provides a number of tools for describing the structure of multimedia content in time and space. The Segment DS describes a spatial and/or temporal fragment of multimedia content – a number of specialized subclasses corresponding to the specific types of multimedia segments, such as video segments, moving regions, still regions and mosaics, which result from spatial, temporal and spatio-temporal segmentation of the different multimedia content types are derived from it. Properties are also associated with the segment classes. For example, we can use properties to define the location of a segment within its containing media object. These include properties such as mediaLocator, spatialLocator, mediaTime (temporal locator) and spatioTemporalLocator. Certain low-level visual and audio features can be associated with different segment types, but not all properties are applicable to all types of segments. MPEG-7 also supports a number of basic non-multimedia entities, which are used in different contexts across MPEG-7, like agent, role and so on. Since MPEG-7 is defined in terms of an XML Schema, the semantics of its elements have no formal basis. Intended meaning is associated with the XML Schema but XML Schema, although flexible, cannot specify semantics in an explicit fashion. In order to overcome this problem, a number of ontology specifications have been proposed. In this section we highlight key features of these ontologies.

Given the MPEG-7 class hierarchy and property specifications, RDF Schema representation for the same can be generated to capture class and property hierarchies. RDF Schema for MPEG-7 representation defines the data model. RDF Schema can specify the constraints on the property-to-entity relationships. Effectively, in terms of the applicability constraint of the properties, RDF Schema makes semantics of MPEG-7 DS explicit. We illustrate this with the help of an example. Consider the MPEG-7 specified descriptors for color feature listed in table 4.1. RDF Schema is able to express constraints of the color descriptors with reference to segment definitions of MPEG-7, through the domain and range values in the color property definitions as shown in figure 4.4.

The above-mentioned example illustrates use of the RDF Schema for ensuring correct usage of MPEG-7 descriptors. This is necessary because the MPEG-7 description scheme – being simple XML Schema – cannot, on its own, represent and check the correct use of the description tools according to their informal yet intended semantics. Ontologies defined with MPEG-7 descriptions can be used to express constraints and can use inference tools to check the semantic consistency of the descriptions. The ontology language OWL provides a more powerful and generic tool to capture semantic constraints that cannot be handled by pure XML Schema. Using OWL, one can define

Feature	Descriptors
Color	DominantColor ScalableColor ColorLayout ColorStructure GoFGoPColor

TABLE 4.1: MPEG7 Color Descriptors

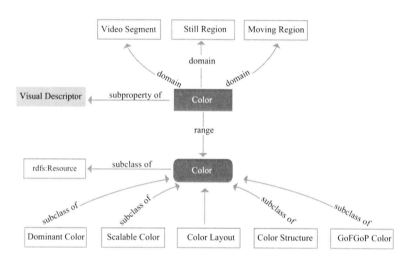

FIGURE 4.4: RDF Class and Property Representation of the MPEG-7 Color Descriptor [101]

an ontology containing the concepts and properties necessary to represent some structural restrictions. RDF statements assert the class membership of particular description tools. For example, both `GlobalTransition` and `Shot` classes are defined as a subclass of a VideoSegment in an ontology for video. The class `KeyFrame` is also defined as a `VideoSegment` but with the semantic constraint that it cannot be further decomposed temporally and that it must be contained in a `Shot` (see figure 4.5). The functional role of a particular type of `VideoSegment` can be inferred by an OWL reasoner. In addition, rules can also be used for specifying implicit constraints because OWL expressivity is not enough for capturing all the semantics constraints. For example, in the context of video segments, a rule can check whether the start/end time of the segments involved in a temporal decomposition are compatible with the start/end time range of their encompassing segment [191].

```
Class(a:Segment) Class(a:VideoSegment partial
  a:Segment
  restriction(a:hasStructuralUnit cardinality(1)))
Class(a:VideoProgrammeSegment complete
  a:VideoSegment
  restriction(a:hasStructuralUnit value (
    mpeg7:urn_x-MPEG-7-davp_cs_StructuralUnitCS_2005_vis.programme)))
Class(a:GlobalTransition complete
  a:VideoSegment
  restriction(a:hasParent allValuesFrom(a:VideoProgrammeSegment))
  restriction(a:hasStructuralUnit value (
    mpeg7:urn_x-MPEG-7-davp_cs_StructuralUnitCS_2005_vis.transition)))
Class(a:Shot complete
  a:VideoSegment
  restriction(a:hasKeyframe minCardinality(1))
  restriction(a:hasParent allValuesFrom(a:VideoProgrammeSegment))
  restriction(a:hasStructuralUnit value (
    mpeg7:urn_x-MPEG-7-davp_cs_StructuralUnitCS_2005_vis.shot))))
Class(a:Keyframe complete
  a:VideoSegment
  restriction(a:hasParent allValuesFrom(a:Shot))
  restriction(a:hasTemporalDecomposition maxCardinality(0))
  restriction(a:hasStructuralUnit value (
    mpeg7:urn_x-MPEG-7-davp_cs_StructuralUnitCS_2005_vis.keyframe))))
```

FIGURE 4.5: Excerpt of the OWL ontology for `VideoSegment` [191].

A refinement of the scheme discussed here was proposed in the BOEMIE approach [43]. An important contribution of this approach was the introduction of the concepts of `SingleMediaItem` and `MultipleMediaItem`. These concepts were introduced to discriminate between items consisting of a single media item and composite ones that contain multiple different media content items (for example, an image with a caption is defined as a type of composite content that includes an image and associated text). Axioms have been included to formally capture the definitions of the composite items. A model for grouping different types of multimedia segments have been formulated based upon decomposition dimensions. These are:

- Spatial, where only position information is used to identify the desired segment

- Temporal, where time instance or time interval is used to define the corresponding segment

- Spatio-temporal, where both position and time related information is used.

The concept `SegmentLocator` is defined to identify a particular segment (spatial, such as a visual mask, and temporal, such as a time interval). A number of additional properties have also been defined. These include the association of an item with the Uniform Resource Locator (URL) providing the physical location of the file and the association of a segment/content item to the domain concepts it depicts. Another additional feature is the support for specifically representing cases where content of one modality is rendered through another – for example, textual information displayed on athletes shirt rendered as a still region of an image.

Rhizomik ontology [43], developed within the ReDeFer5 project, is based upon a fully automatic translation of the complete MPEG-7 Schema to OWL. Therefore, it works with OWL DL ontology covering all elements of the entire MPEG-7 standard. The Core Ontology for MultiMedia (COMM) [181] is in OWL DL and covers selected tools from the structural, localization and media description schemes, as well as low-level descriptors of the visual part. Additionally, unlike other approaches, it provides the means to include in the annotations information about the algorithm and corresponding parameters used in the extraction of a given description.

Ontology specifying semantics of MPEG-7 descriptions has to be combined with domain knowledge so that applications can take advantage of the additional knowledge. We examine how MPEG-7 descriptions can be combined with domain knowledge for developing applications in the next section.

4.4   Using MPEG-7 Ontology for Applications

MPEG-7 ontology, as discussed in the previous section, has limitations in semantic processing of multimedia content. Consider as an example the query discussed in Tsinaraki et al [192]. The query is for retrieving the audiovisual segments where the soccer player Zidane scores against the goalkeeper Buffon. Using only the MPEG-7 constructs, the exact query cannot be constructed. Using the definition of an MPEG-7 event involving agent and experiencer, query can be posed as instead: Give me the segments where an event takes place, having Zidane as agent and Buffon as experiencer. This query will result in retrieving segments with Zidane and Buffon being the agent and the experiencer, respectively, of any event, and those segments need not be the video of the goal-scoring event. Using domain knowledge captured in soccer ontology (which provides a conceptual model of the goal event) together with the MPEG-7 constructs a user can express the query: Give me the segments where a goal event takes place with the player Zidane scoring against the goalkeeper Buffon. This query can be interpreted correctly with the help of domain ontology to retrieve relevant video segments.

4.4.1 Museum Ontology

We discuss how multimedia ontology can be created for multimedia or audiovisual content within museums [99] in this section. Museum collections are highly diverse. Artifacts are distinguished based upon their origin, genre, media type, format, quality, age and context. The metadata for multimedia resources may need to include bibliographic information, description of media-specific details (format, structural or segmentation information) and semantic information (description of the objects/people/places/events recorded). The "CIDOC object-oriented Conceptual Reference Model" (CRM) [100, 99] was developed by the ICOM/CIDOC Documentation Standards Group to provide an ontology to facilitate the exchange of cultural heritage information. The CIDOC CRM has an object-oriented extensible data model. Figure 4.6 illustrates a class hierarchy for the CIDOC CRM.

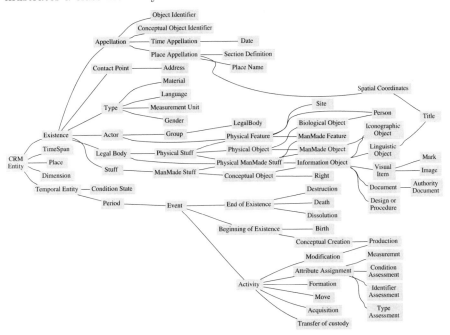

FIGURE 4.6: CIDOC-CRM class hierarchy.

However, the CIDOC CRM is limited in its ability to describe digital multimedia or audiovisual content. It does not have provision for specifying temporal, spatial or spatio-temporal locations within non-physical digital media. It cannot represent visual or audio features such as color histograms, regions, shape, texture, volume, and so forth. Further, there is no provision to specify audiovisual segments, key frames, scene changes, and so forth. CIDOC ROM can be combined with an MPEG-7 based ontology to resolve these problems.

Hunter [99] proposes a scheme for enhancing CIDOC ROM ontology. The CIDOC CRM provides an `is_composed_of` property that can be used to define the structural or segmentation metadata associated with multimedia resources. Spatial, temporal, spatio-temporal and media-source decompositions are all provided through this approach. Visual and audio descriptors are provided through the provision of two new classes, `VisualFeature` and `AudioFeature`. The properties `has_visual_feature` and `has_audio_feature` are used to associate these new classes/subclasses with the relevant multimedia document types.

4.4.2 Associating Domain Knowledge with MPEG-7 Ontology

Rhizomik ontology [43], introduced earlier, covers the Semantic DS description tools of MPEG-7. Semantic content is represented through a set of subclasses of `SemanticBaseType` – namely, the `SemanticType`, `AgentObjectType`, `ObjectType`, `SemanticTimeType`, `SemanticPlaceType`, `SemanticStateType`, `ConceptType`, and `EventType` classes. This abstraction model defines a rather coarse conceptualization. This approach is applicable only under the presumption that domain ontologies are compliant to the subclassing scheme. Obviously this makes linking of metadata with existing domain ontologies a major challenge as it requires tedious mappings that cannot be easily automated. Applications have been designed for structured domains with this approach. Application examples include the semantic integration and retrieval of music metadata.

Context [104] can be used to add a semantic layer on top of the existing low-level feature-oriented multimedia representation of MPEG-7 to facilitate high-level semantic integration. Context layer is useful in identifying related contextual information for a clearer semantic meaning. For example, the word *kick* has no clear semantic meaning but *free-kick*, *Beckhams free-kick*, or *David Beckhams free-kick at the FIFA World Cup 2006 in Germany* have clearer semantics [104]. Hence, semantics of a concept becomes clearer with more contextual information. This additional information is referred to as contextual information [54]. Therefore, multimedia information retrieval systems need to utilize this contextual information. Typical multimedia contextual sources include semantic annotation of multimedia resources: image, audio, video; knowledge sources, such as information from transcription of the audio part of video files; entities' properties and general descriptive metadata, such as author, title, date of publication; and inference and deductions based on a well-defined ontology.

Ontologies based on MPEG-7 describe the multimedia content at a number of levels and provide a way to describe what is depicted in the multimedia content from the real world, such as objects, people, and events. A context-based ontology model of multimedia content facilitates disambiguation and specification of clearer semantics. A formalized model, presented in [54], has been used to generate RDF-OWL context-based multimedia semantics.

4.4.3   Ontological Framework for Application Support

The development of ontology guided multimedia applications requires information flow and integration among different knowledge sources and processing elements. The DS-MIRF framework [192] aims to support the development of knowledge-based multimedia applications utilizing and extending MPEG-7/MPEG-21 standards. The architecture of the DS-MIRF framework along with the information flow between its components and the interaction with end-users is presented in figure 4.7. This shows the process of OWL-based semantic annotation of multimedia content using the ontological infrastructure of the DS-MIRF framework. An annotator uses a specialized annotation interface. The OWL descriptions are then transformed, using the appropriate set of transformation rules, to standard MPEG-7/MPEG-21 metadata descriptions. These are stored in the DS- MIRF MPEG-7/MPEG-21 metadata repository, which is accessed by end users through appropriate application interfaces. The application interfaces may provide the end users with multimedia content services like multimedia content retrieval or filtering. An integrated OWL reasoning engine helps in complex constraint checking, logic-based knowledge and complex query processing. The validation of both ontologies and ontology-based metadata takes place during the annotation process.

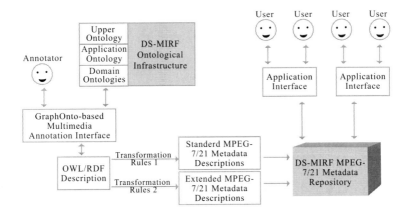

FIGURE 4.7: Information flow in the DS-MIRF framework.

Ontologies are organized as shown in figure 4.8 in DS-MIRF. An OWL upper ontology captures the MPEG-7 Multimedia Descriptor Schemes (MDS) and the MPEG-21 Digital Item Adaptation (DIA) architecture. It includes OWL Application Ontologies, which provide an additional functionality in OWL that either makes it easier for the user to use MPEG-7/MPEG-21 or supports advanced multimedia content services, such as user preferences. Domain ontologies encode domain knowledge. For example, sports ontologies extend the abstract semantic description capabilities of the MPEG-7 MDS in the domain of sports like football. In the domain of football, domain ontology specifies the concept of goal, expressed in OWL. A video-retrieval query for frames depicting goal makes use of the definition of the concept of goal defined in the domain ontology.

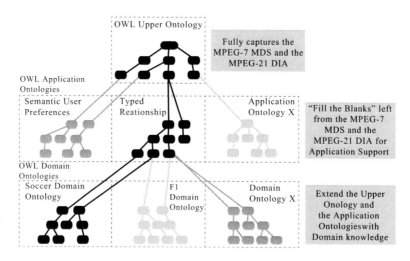

FIGURE 4.8: The ontological infrastructure of the DS-MIRF framework.

4.4.4   MPEG-7 and Semantic InterOperability

Semantic interoperability is the ability to automatically process the web-based information by agents, querying engines and so on. Multimedia information expressed in MPEG-7 XML Schema can be exchanged between processing agents due to the accepted syntactical format of the standard but the task of implementation of the intended semantics is assigned to the applications. As an alternative, each multimedia repository can provide a formal description in an ontological form by adopting the most appropriate MPEG-7 ontology. Other multimedia repositories and applications can exploit the meaning encoded in the ontology, provided automated mapping and reasoning tools can establish desired alignment between the ontologies.

Another approach haqs been suggested to address this problem of automated mapping generation. Different multimedia repositories can use their own multimedia ontologies and also define mappings (manually or automatically) to core and domain multimedia ontologies, which may be globally available or among which appropriate mappings/alignments already exist. A core ontology is a complete and extensible ontology that expresses the basic concepts that are common across a variety of domains and media types. These concepts can provide the basis for specialization into domain-specific concepts and vocabularies. CIDC-ROM, discussed earlier, is an example of a core ontology related to museum artifacts. A more general top-level core ontology ABC is proposed in [100] that enables domain-specific ontologies to be incorporated, as they are required. It has been encoded as a set of RDF Schemas. The ABC ontology can model physical, digital and analog objects held in libraries, archives, and museums and on the Internet. The ontology can model objects of all media types: text, image, video, audio, web pages, and multimedia. It can also represent abstract concepts. In addition, the model can be used to describe other fundamental entities that occur across many domains, such as agents (people, organizations, instruments), places and times. The top-level class hierarchy of ABC ontology is presented in figure 4.9.

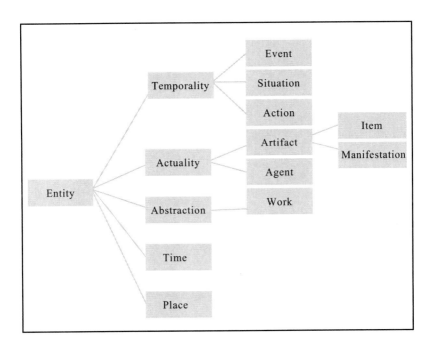

FIGURE 4.9: ABC ontology.

The core of the ABC models are the entities. The `Temporality` class of ABC model provides a unique mechanism to model time and the way in which properties of objects are transformed over time. A Situation provides the context for framing time-dependent properties of (possibly multiple) entities. An Event marks a transition from one situation to another. An Action provides the mechanism for modeling the knowledge about the involvement and responsibility of agents in events. The Actuality ontology category encompasses entities that are sensible – they can be heard, seen, smelled, or touched. This contrasts with the Abstraction category, which encompasses concepts. These top-level sets of classes of ABC and their properties can act as attachment points for domain-specific ontologies. In the case of the MPEG-7 ontology, the obvious attachment point for the MPEG-7 class hierarchy is the ABC Manifestation class. For example, a video-segment is an instance of manifestation class, which by virtue of temporality will be associated with an event and thereby linked to domain-specific metadata of the event. This is illustrated in figure 4.10. Multimedia data from different repositories can, therefore, be linked to the domain ontology of the event with the help of ABC core ontology.

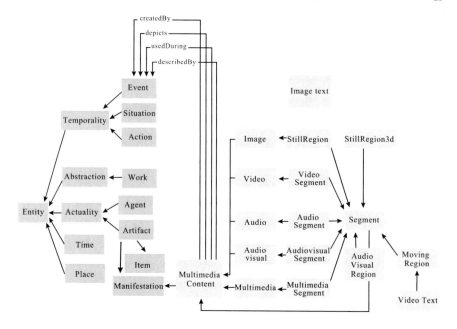

FIGURE 4.10: Linking ABC ontology with MPEG-7.

4.5   Multimedia Concept Modeling

In the previous section, we described mechanisms developed for linking domain ontology with ontologies built upon MPEG-7 description standards. However, we have not examined the issue of concept modeling in the multimedia domain. Although MPEG-7 specifies a large number of classification schemes for multimedia content, it has been a challenge to develop tools for automated extraction of these descriptors for annotation. However, we need to have automated tools for objective conceptual annotation. These tools are required to build large-scale concept ontology facilitating end-user access to multimedia repositories and covering the semantic space interesting to end users.

Large Scale Concept Ontology for Multimedia (LSCOM) [138] is one such initiative. The LSCOM effort selected broadcast news video as its target domain due to the significant interest in news video as an important multimedia source of information and has produced an ontology of more than 1000 concepts. A set of concepts, with reference to the news corpus, was identified through an annotation process. Out of these concepts, those observed in the corpus – such as airplane flying, protests, riots, parade, and aircraft carrier – were deemed feasible for automated tagging. On the other hand, more subjective concepts (such as discovery, challenge, and happiness) were identified as unobservable and infeasible. Overall, candidate concepts were filtered down to 834 based on feasibility and utility. To create a fully annotated video dataset, the LSCOM team labeled the presence or absence of each of the 449 concepts in 61,901 shots of broadcast news video. This required more than 28 million human judgments. Labeled examples of concepts in the video data were used to learn classifiers for detection of concepts in the video. The taxonomy design organized the 834 concepts into six categories on a top level: objects, activities/events, scenes/locations, people, graphics, and program categories. These categories were further subdivided, such as, by subdividing objects into buildings, ground vehicles, flying objects, and so forth. Appropriate mechanism was designed to represent the hierarchy of LSCOM concepts using OWL. The LSCOM ontology also includes most binary relations known to hold among target concepts, as well as higher relations (so-called rule macros) that stand in for rules by relating target concepts to binary relations. This also includes many explicit rules (universally quantified if/then statements), so long as they can be represented in Semantic Web rule language. Following another approach, Bertini et al [17] proposed an ontology representation scheme, which included linguistic concepts and visual prototypes to encode manifestations in the visual world. These prototypes were obtained by clustering the instances of visual data that are observed using appropriate features. This framework was used for annotating video streams.

4.6   Ontology Applications

Ontology is a critical component for a variety of multimedia-centric applications. We have discussed various generic technological issues in the previous sections. In this section, we consider some domain specific ontologies and their applications. The examples described here are representative and do not present an exhaustive account.

4.6.1 Use of Ontology for Accessing Paintings

Multimedia information retrieval in knowledge-rich domains like fine arts, architecture and performing arts poses challenges that are different from other domains. The media-dependent and media-independent information are to be encoded in a form that reflects the semantics of the domains' relationships. Context and background information influence the perception of works of art. Ontology helps in organizing this knowledge by associating appropriate background information.

Vrochidis et al. combined ontology-based and visual feature-based retrieval for heritage images [200]. A user could pose a query by specifying example images or by specifying constraints on annotation. Annotations are linked with ontology concepts. Features extracted from query images are compared against precomputed similar features from images in the heritage collections. The ontology-based search makes use of custom-built ontologies specific to the annotation structure of the individual collections. The system recommends images based upon content-based similarity and/or satisfaction of ontological constraints. To make the system interoperable with other ontologies in the cultural domain, the custom-built ontologies are mapped into the CIDOC CRM.

Another ontology guided painting retrieval and browsing system has been presented in [103]. The system organizes paintings according to the metadata associated with such works. The metadata include details of the artist, the artists' dates/places of birth (and death), dates when and where the image was created, influences on the artist, and relationships (if any) to other images. The terminology and named entities associated with an image are organized in a hierarchy of object names and other relationship graphs, like graphs encoding meronymic (part-of) relations. Interesting aspects of this system are its structured organization of ontology and knowledge base and its design of an ontology-guided query engine. The architecture of the system is shown in figure 4.11. The system has three components: (i) ART ontology and knowledge base; (ii) Artfinder : a visual query-processing interface; and (iii) an ontology populator to automatically build up the ontology. The entire system is implemented in OWL.

The key component of the system is the ontology and its knowledge base. The ontology structures information about a domain, and the knowledge base is the data belonging to that domain. In terms of description logic terminology, parts of the A-Box shown in the figure is considered as belonging to the ontology. This refers in particular to art historical periods and geographical entities. In accordance with the modeling principles of the CRM and other, general ontological considerations, art historical periods are modeled as instances and are, therefore, part of the ABox. Similarly, nearly all the geographical information in our system is contained in the A-Box. This kind of information has a more generic character than assertions on individual works of art.

The query interface had two sections. First, a user can formulate his or her

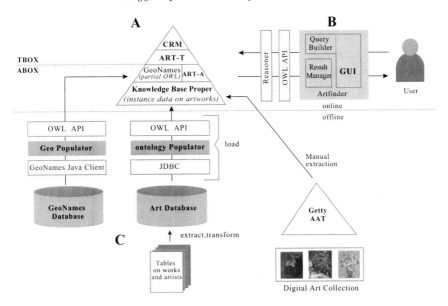

FIGURE 4.11: Architecture of a painting browsing system.

query using several input fields and selection menus. Second, a table displays the results of a query: images and basic information about works of art matching the specified criteria. Graphic User Interface (GUI) for query formulation is designed in terms of segments as tabs, representing classes in ontology, and these segments are linked through interface elements to object properties in the ontology. Labeled buttons are used to represent ontology properties.

The systems discussed above illustrate the process of designing an ontology-based heritage object retrieval system for knowledge-rich environments. Query interface also provides novel GUI design for enabling users to use ontology for formulating queries for multimedia artifacts like paintings in a flexible manner. However, these approaches have not considered exploitation of ontology for interpreting multimedia content.

4.6.2 Ontology for Ambient Intelligence

Ambient intelligence is defined as a digital and proactive environment with the capacity to sense the environment and assist users in their daily lives. Ambient intelligence (AmI) systems are usually composed of (a) a sensing mechanism to gather information from both the user and the environment, (b) a set of actuators to modify the environment and communicate with users, and (c) a reasoning/decision-making module. The reasoning module should have the ability to recognize what is happening to users in the environment, what they do, what their aims are and how to make decisions to assist them. The decision module interprets input from different sensors, including a camera, to infer the state of the users and their context. Ontology in this application provides a common vocabulary to interpret scenarios. Further, ontology provides a mechanism to reason about scenarios. This ontology is typically encoded in a standardized ontology language like OWL. For example, ontology can be used to define a `FriendlyGathering` as an activity having at least two actors who are friends. Reasoning capability can be used to derive implicit information from explicit context data. For instance, the users' current activity can be inferred based on his or her current location, posture, used objects, and surrounding people. A simple descriptive condition – Two people working in the same project and institution are colleagues – may be formulated to infer which people are colleagues in the system based upon description logics in OWL2 [165] in the following way:

```
Person(?x) ∧ Person(?y)
    ∧worksInProject(?x,?p) ∧ worksInProject(?y,?p)
    ∧worksForInstitution(?x,?i) ∧ worksForInstitution(?y,?i)
    → isColleagueWith (?x,?y)
```

When conceptual models of the activities and their interrelationships are encoded in the ontology in a machine-understandable way, a higher degree of automation and flexibility is possible. An approach based exclusively on ontology reasoning [32] uses ontologies to represent activities as well as each data source that can be used to recognize them, from sensors to actors. Coarse-grained activities are recognized by ontological reasoning based on the available data, which are refined as new information becomes available.

As new data sources are introduced in the environment, pure data-driven learning-based approaches require retraining of the complete module. Furthermore, to rebuild the updated model, data collection with the new data sources is required. However, use of ontology allows previous activity-recognition models to be reused; updating only the affected condition is sufficient for recognizing the adapted activity. Despite these advantages, a major limitation in a pure ontology based approach is the lack of support for imperfect information.

As an example, we consider the human activity recognition ontology proposed in [164], which models individual and social activities. The types of interaction are modeled as acknowledgment, asking for opinion, comment, negative/positive opinion, proposal, or request of information. Activity granularity is specified only at one level. However, an extensive taxonomy is created for personal, physical, professional activities, traveling activities, postures and activities using different kinds of devices or artifacts. An interval-based representation models the overlapping of activities in time. Other entities are indoor (corridor, arcade, elevator, etc.) and outdoor (promenade, sidewalk, rails, etc.) communication routes. Symbolic locations (indoor, outdoor, pedestrian or not) are also supported. The system maps context data to ontological classes and uses a hybrid ontological/statistical reasoner to retrieve information about simple human activities. Ontological reasoning with OWL-2 is used to recognize complex activities based on elementary sensor data. Simple activities are recognized through data-driven methods like a simple Bayes classifier.

In general, expressiveness requirements in human behavior and environment representation include the ability to represent hierarchical structures, complex relationships among context instances and complex definitions based on simpler ones, usually using restrictions that may be spatial or temporal. Discussions in this section have shown that ontologies provide one of the most promising tools to achieve these goals. Further, ontology-based reasoning provides an efficient mechanism for interpreting data and contexts generated from multiple sensors, each of which can be sensing different signals.

4.6.3   Ontology for Sensor-Web Applications

Observations, and the sensors that obtain them, are at the core of a number of data-centric applications ranging from meteorology to medical care to environmental monitoring to security and surveillance. In a generic sense these sensors provide multimedia data, which, in effect, in many cases goes beyond the scope of the multimedia description scheme of MPEG-7. Ontological models may allow autonomous or semi-autonomous agents to assist in collecting, processing, reasoning about, and acting on sensors and their observations. A standardized ontology in OWL-2 has been proposed in [36]. The ontology describes sensors, the accuracy and capabilities of such sensors, observations and methods used for sensing. Concepts for operating and survival ranges are also included, as these are often part of a given specification for a sensor, along with its performance within those ranges. Finally, a structure for field deployments is included to describe deployment lifetime and sensing purpose of the deployed instrument. The ontology can be seen from four main perspectives:

- A sensor perspective, with a focus on what it senses, how it senses, and what is sensed;

- An observation perspective, with a focus on observation data and related metadata;

- A system perspective, with a focus on systems of sensors and deployments; and,

- A feature and property perspective, focusing on what senses a particular property or what observations have been made about a property.

The ontology is used in the Grid4Env project [73, 74], which aims to build large semantic sensor network (SSN) applications for environmental management. The SSN ontology is linked with ontologies for coastal features, services and roles for emergency response. These applications requiring ontology at various levels are expected to grow rapidly in number. Engineering ontology based Semantic Web solutions for a large number of sensors (maybe millions) and their multimedia observations is an interesting research problem.

4.6.4   Biomedical Applications

Biomedical ontologies have played an important role in the knowledge management tasks and in annotation processes. Foundational Model of Anatomy (FMA) ontology is an example that is considered a reference model about anatomy. This ontology represents in a symbolic way the structure of the human body from a macromolecular to macroscopic level [229]. It was conceived as a general-purpose resource to be used in different problems requiring anatomical knowledge [229, 89]. The FMA was implemented in Protege-Frames and stored in a MySQL database. It contains approximately 75,000 classes and over 2.1 million relationship instances. These ontologies can be used in a variety of applications including image annotation.

The annotation/metadata associated with images can be classified as:

- Content-independent metadata, where metadata are related to the image but not linked with the content, such as names of authors, dates, location, and so forth

- Content-dependent metadata, where metadata is related to low-level and/or intermediate level features,such as color, texture, shape

- Descriptive content metadata, where metadata is related to semantic content. The descriptive content metadata can be provided at two levels of specificity:

  - Descriptive content associated with the full image
  - Descriptions linked with descriptive content in each segmented region.

  Systems annotate images with descriptive content using metadata based on ontologies.

Ontologies are used typically for two purposes, one to define an annotation scheme and another to define domain concepts. For example, a structured set of object properties defined in the ontology can be used to specify the annotation scheme. Ontologies, for image annotation, define a set of visual entities and map the appearance of these entities to their textual descriptions. This association creates a bridge between the visual characteristics of important regions within an image and their semantic interpretation. For example, an area of a CT scan having a slightly bright and hazy appearance (a visual entity) might be mapped to "ground-glass opacity" (its textual description). Second, a visual ontology defines relationships among the entities. If the ontology maps visual entities to concepts within existing textual ontologies, the relationships among visual entities can be aligned with those of their corresponding textual entities. A content-based image retrieval system can extract important regions from an example query image, map the visual characteristics of these regions to textual concepts, and then use these concepts to search images in the image collections using annotations associated with images.

4.6.5 Ontology for Phenomics

An organisms phenotype is an observable or quantifiable trait of the organism. Phenotypical variations happen as a consequence of its genetic makeup combined with its developmental stage, environment and disease conditions. Phenomics is the systematic and comprehensive study of an organisms phenotype based upon a huge volume of data generated from high-throughput and high-resolution imaging and measurement-based analysis platforms. Use of ontology in developing a data repository for large-scale phenomics data (PODD: Phenomics Ontology Driven Data repository [122]) can ensure efficient and flexible repository functionalities. PODD provides a mechanism for maintaining structured and precise metadata around the raw data so that they can be distributed and published in a reusable fashion. The domain models have been constructed in OWL. The OWL domain model helps in the creation, storage, validation, query and search of data and metadata. Domain concepts are modeled as OWL classes; relationships between concepts and object attributes are modeled as OWL object and data type predicates. Concrete objects are modeled as OWL individuals.

The top-level domain concept in PODD is "Project". Another domain concept is "Technical measurement platform" – that is any platform for which parameters and parameter values may be captured – used in the project. Another domain concept that is of central importance is "Investigation", which effectively captures different aspects of experimental investigation. It captures the data and metadata of experiments under a project. This application (PODD) illustrates how ontology can be used for management and distribution of multimodality data.

4.7 Conclusions

We discussed multimedia ontologies and use of ontology for management of multimedia information in this chapter. We saw that Semantic Web technologies like RDF and OWL have been used for specifying multimedia ontology and domain ontologies. Intended semantics of MPEG-7 description schemes have been made explicit using RDF-based and OWL-based ontologies. Domain ontologies expressed in OWL have been linked with MPEG-7 ontologies for specifying domain-linked semantics. Multimedia information is in general associated with content-based and content-independent metadata. OWL-based ontology has provided the platform for representation and reasoning with the domain model for the metadata for different applications. Although LSCOM provided a taxonomy of multimedia concepts but did not provide a generic framework for defining new multimedia concepts. A basic problem with OWL-based ontology is the inherent incapability of OWL to handle incompleteness and uncertainty in the domain. Further, OWL-based specification of conceptualizations cannot link concepts with their manifestations in the media or signal space in terms of explicit observation models of concepts. In the next chapter, we present a new ontology specification scheme that addresses these issues.

Chapter 5

Multimedia Web Ontology Language

5.1 Introduction

The ontology languages discussed in this book so far represent a conceptual model for a domain. Multimedia instances represent a perceptual recording of the world. There is often a mismatch between the sensory inputs and the concepts that they represent. This makes application of ontology for multimedia document interpretation a difficult task. The approaches reviewed in the previous chapter attempt to combine domain knowledge with knowledge about multimedia data representation for semantic access to multimedia. However, they fail to address the inherent uncertainties that are associated with media manifestations of the events and the incompleteness associated with the observations that are common to the multimedia world. In this chapter, we introduce a new ontology representation scheme, Multimedia Web Ontology Language (MOWL), that addresses these limitations.

We have organized this chapter as follows. We introduce perceptual modeling of domains that complements traditional conceptual modeling in section 5.2. The following section (section 5.3) presents a brief overview of the ontology representation scheme MOWL, which is based on perceptual domain models. The following sections (sections 5.4 and 5.5) introduce the formalisms in the ontology representation scheme. The language constructs of MOWL are described in section 5.6. Section 5.7 provides a formal description of the Observation Model(OM) which is a key contribution of MOWL and is used for concept recognition. The inferencing mechanisms for the construction of OM and concept recognition are described in section 5.8. Section 5.9 discussed possible other inferencing schemes with MOWL. Finally, we analyze the strengths and weaknesses of the ontology representation scheme in section 5.10. Formal definitions of the language schema, MOWL semantics and examples of media property specifications can be found in the appendices at the end of this book.

5.2   Perceptual Modeling of Domains

As computer vision researchers struggle with resolving the *semantic gap* that exists between sensory inputs and the concepts that they represent, human minds are engaged in concept recognition tasks myriads of time during daily activities. They perform the tasks successfully (at least most of the times) without much apparent effort. Indeed, concept recognition tasks appear to be so trivial that we "see" a car moving or an airplane flying, while what we really see is some visual patterns appearing on the retina of our eyes. This motivates us to understand and replicate the working of the human mind in relating the concepts with the percepts.

A closer look into the human cognition system reveals that the *semantic gap* between percepts and concepts is rather artificial. From the discussions in chapter 3, we recall that the concepts are nothing but a symbolic representation of the percepts in their abstract form. An observation of the real world object or event results in reception of sensory inputs that create a mental model of the concept in terms of perceptual patterns. For example, when a person observes a steam locomotive, his mental model for the concept may comprise visual attributes such as the body shape of the locomotive, motion attributes such as the movement of the pistons and the wheels, and audio attributes, such as its whistle and huff-and-puff. Similarly, observation of a dance performance results in creation of a mental model comprising the dress of the artiste, her body movements and the accompanying music. Observation of many instances of objects and events results in grouping of such perceptual inputs into different conceptual classes, which are then expressed with a linguistic pattern, such as "steam locomotive" or "classical dance". Analysis of similarities and dissimilarities of these media features of different objects and events gives rise to concept hierarchy. Further, the observations often get associated with each other — for example, the body shape of a steam locomotive with a cloud of smoke or a pair of straight lines, that is, the rails. Analysis of such perceptual associations reflects in semantic associations across the corresponding entities in the conceptual space.

The natural representation of concepts in the human mind remains in the form of a set of correlated perceptual media features, which are abstractions over many observations of the concept. As a result, when we "look for" an instance of a concept, we indeed look for the perceptual properties signified by the concepts in the perceptual world, for example, in an image or a video. Even in a text document, we look for the visual pattern that represents the symbolic representation of the concept. We recognize the concept when we observe some of these expected media patterns. For example, an Indian monument of the medieval era is recognized by its characteristic domes and minarets, its wide facade and similar other properties. Figure 5.1 depicts perceptual modeling of the concept `Steam Locomotive`.

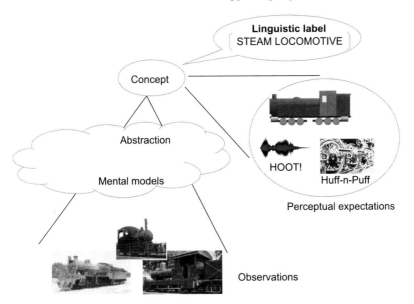

FIGURE 5.1: Illustrating perceptual modeling of a concept (images are borrowed from Wikimedia Commons).

We observe that a perceptual model, comprising a set of media patterns, represents a manifestation of a real-world concept in media documents. Observation of some of these media patterns in a media form is evidence of the existence of the concept. Thus it is natural to model the association between the concepts and their media properties as a *causal model*, where the abstract concepts are the causes for a set of media features that can be perceived with our sensory organs. The concept recognition takes place through an abductive (evidential) reasoning mode when some of these media patterns are observed. Needless to say, there are large variations in the media patterns that represent the manifestations of the concepts and a concept may often manifest in multiple alternate perceptual models.

Although it is possible to associate media patterns with concepts in the OWL/RDF-based representation of domain models, the semantics of the representation scheme do not signify the causal model. The DL-based reasoning scheme supported with OWL does not permit the evidential reasoning that is useful for concept recognition. In the following sections of this chapter, we introduce a new ontology representation scheme that is based on this causal model of the world and supports abductive reasoning. Several applications of this ontology representation scheme follow in the subsequent chapters.

5.3   An Overview of the Multimedia Web Ontology Language

A new ontology representation scheme, Multimedia Web Ontology Language (MOWL) [131], is designed for perceptual modeling of a domain. The ontology representation scheme proposes a causal model of the world, where the concepts manifest into media patterns in the multimedia documents. This causal model is complemented with an abductive reasoning system for concept recognition with evidential strength of the media patterns observed in multimedia documents.

5.3.1   Knowledge Representation

In the domain representation scheme of MOWL, media properties are associated with concepts using a special type of property, which signifies causal association. These associations are characterized with uncertainties because of inherent differences across the concept instances and their media manifestations. These uncertainties have been expressed using Conditional Probability Tables (CPTs) as in [48]. However instead of using the probabilities directly for belief propagation, a Bayesian network derived from the ontology is used for inferencing. The Bayesian network incorporates domain-specific contextual information as explained in the following text.

Context plays an important role in concept recognition. For example, observation of an Indian flag reinforces the belief in sighting Narendra Modi (prime minister of India), even when his face is not recognized with adequate confidence. Similarly, a specific type of music may reinforce the belief in some specific dance form, and vice versa. Thus, a perceptual model for a concept "inherits" media properties of related concepts. To explore the nature of such property "inheritance," consider an ontology on heritage monuments that may encode the knowledge that the Taj Mahal is an instance of "Indian medieval monuments" and that it is built with "marble." This background knowledge suggests that a visual portrayal of the Taj Mahal is likely to exhibit the general structural composition of Indian medieval monuments, as well as the color and texture properties of the stone. On the other hand, the Taj Mahal being a typical instance of a tomb from the Mughal era, an example image of the Taj Mahal is a valid example for a tomb of that period. Thus, media properties *propagate* across related concept nodes associated in certain ways in a multimedia ontology as shown in figure 5.2. This is a unique requirement for multimedia data interpretation and is quite distinct from the property inheritance rule implied by hierarchical relationship in a traditional ontology. A subclass of properties has been defined in MOWL to indicate such relations. It is possible to reason with the relations defined in a MOWL ontology to derive a Bayesian network that relates concepts with their possible media manifes-

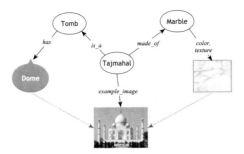

(a) Media property propagation

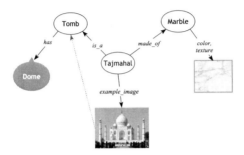

(b) Media example propagation

FIGURE 5.2: Media property and example propagation across related concepts (image of the monument is borrowed from Wikimedia Commons).

tations. This derived Bayesian network is called an *Observation Model* and is used for recognition and association of concepts. The observation model for a concept includes media patterns propagated from the related concepts in the domain. Thus, the observation model represents a media-based description of a concept that incorporates domain-specific contextual information as well.

A further requirement for the multimedia ontology representation language is that it should be possible to represent the media properties of concepts in different ways and at different levels of abstractions. For example, the shape of a dome can be specified either with an abstract geometric description or with a set of example images. A dance posture may best be specified by the binary output of a specialized feature classifier. A multimedia ontology language needs to provide distinct semantics for such variations. Some of the complex media patterns characterizing a concept can be composed of simpler patterns arranged in space and time. For example, the event class "GoalScore" in a soccer match can be characterized by a temporal sequence of audiovisual units, where each of the visuals is a spatial composition of constituent concepts, such as a ball, the players and a goalpost. Like the media properties,

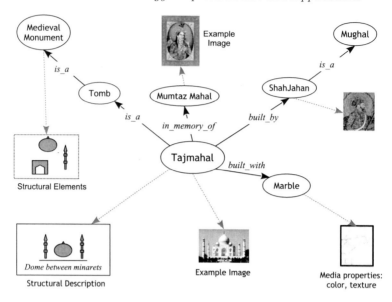

FIGURE 5.3: A section of ontology encoding perceptual properties (images are borrowed from Wikimedia Commons).

the spatio-temporal relations for an event class are marked with uncertainties. A multimedia ontology language should be capable of formal yet flexible description of the structural composition of complex concepts. MOWL adopts the representation for spatio-temporal relations as proposed in [214], which provides mechanisms for describing a complex class of events in terms of elementary concepts and their spatio-temporal relations, which are formally yet flexibly defined. A fuzzy membership function is used to compute the "belongingness" of an event instance in an event class so described. This is a novelty as compared to the *crisp* temporal logic or *informal* MPEG-7 such as the spatial and temporal relations used in some of the earlier works [9, 192].

5.3.2   Reasoning

Concept recognition in MOWL takes the form of abductive reasoning. Concepts are recognized on the basis of evidential strengths of the media patterns observed in a multimedia document. We illustrate the mechanism with an example in this section. Figure 5.3 depicts a small section of an ontology in architecture domain. Figure 5.4 shows a possible OM for a concept, the Tajmahal, derived from that ontology. The OM is organized as a Bayesian tree, with the concept Tajmahal placed at the root node. The expected media patterns, when the monument manifests in a media instance (image or video), appear at the leaf nodes of the tree. Note that the Tajmahal "inherits" the media properties of some related concepts in the domain. The structure of

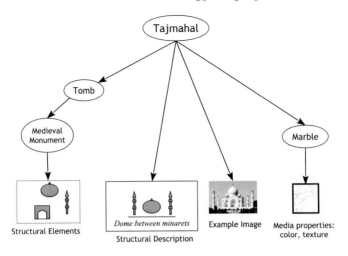

FIGURE 5.4: A possible observation model for the concept `Tajmahal`.

the Taj Mahal has been specified as a spatial composition of its constituent concepts, namely a facade, a dome and the minarets. For recognizing the concept `Tajmahal` in a media document, the Bayesian tree is initialized and appropriate media detectors are run to discover the media patterns specified at the leaf nodes. When a pattern is discovered, the corresponding node is instantiated resulting in a belief revision in the network. The presence of the concept is inferred by virtue of the posterior probability of the root node as the cumulative result of such belief revisions.

5.4 MOWL: Concepts, Media Observables and Media Relations

A concept is an abstraction of some human percepts. From one viewpoint, it represents a class of real-world objects and maps to a "class" in common knowledge description schemes. From another point of view, it represents some expectations for specific sensory patterns. Observation of those patterns leads to recognition of the class of objects that it represents. Thus, the first and foremost requirement for a multimedia knowledge representation scheme is that it should allow for declaring sensory properties of the concepts (classes).

MOWL distinguishes between two types of entities, namely (a) the *concepts* that represent the real-world objects or events and (b) the media properties of the concepts, including *media patterns* and *media instances* that represent the manifestation of concepts in different media forms. Presence of these media patterns and instances in a media document can lead to recognition of the related concept in the document. For example, while a person is a real-world concept, the visual appearance of the person's face is a media pattern, detection of which in an image or a video clip leads to recognition of that person's presence in the real-world event captured in the media clip. As another example, a specific performance of a dance piece can be recognized by a set of gestures, postures and actions, that form a set of media patterns representing the possible media manifestations for the dance performance of a particular conceptual category.

These patterns can be represented in an ontology as descriptions through media content representation standards like MPEG-7 descriptors in different ways, the simplest being a low-level media feature specification possibly using MPEG-7 tools, for example, $\langle Mpeg7 \rangle \ldots \langle ColorSpaceType =$ "RGB"$/\rangle \ldots \langle /Mpeg7 \rangle$, which specifies that the color space used in the image visual feature is RGB. On the other extreme, complex media features, such as a dynamic body posture, generally require dedicated pattern-detection tools. MOWL provides for a media pattern to also be represented as a specification of a module or procedure that has the capability of extracting a media feature from a media resource, comparing it with an expected value of the media feature and returning the similarity computation result as a belief value for the presence of the expected media feature. Any formal procedural specification language can be used to specify such a pattern-detector specification. One such procedural pattern specification is Web Service Definition Language (WSDL) specification of the pattern detector, an example of which is shown in appendix C.

Concepts in MOWL can also have media instances associated with them— for example, a photograph of a dancer; an image of a monument; a video clip showing a human action, say a soccer goal; or an audio recording of a music performance. Media examples can be represented through a URL that points to the media file or as a procedural specification that specifies an *example-based search* paradigm. For instance, a speaker-recognition routine can build speech models of individuals and an average human speech model with labeled examples of many human speech samples and use those models for speaker recognition. Language constructs of MOWL allow specifications for such routines with the media pattern *speech*. Such specifications define how a media resource file may be compared with a media example and return the belief, which denotes how similar they are. The procedural specification allows defining classifiers as well, which can be trained with a large number of media examples.

5.5 MOWL: Spatio-Temporal Constructs for Complex Events

There have been several approaches to characterize an event situated in space and time [175, 163, 170]. In general, an event can be described by four Ws: namely, who, what, where and when. Formal definition of *event classes* in the context of an ontology, however, needs to follow a different approach. An ontology needs to encode the generic characteristics of event classes that are common to many event instances. A complex event depicted in multimedia format is generally composed of several simpler events or elementary concepts, connected through spatial and temporal relations. An often-cited example is the GoalScore event in a soccer match that can be considered as a spatio-temporal composition of simpler events and concepts as follows:

GoalScore = ((Ball *Inside* Goalposts) *FollowedBy* Cheer)

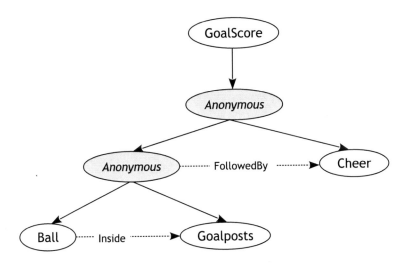

FIGURE 5.5: Spatio-temporal definition for GoalScore event.

This decomposition is schematically shown in figure 5.5. The description is compliant to RDF Schema description. A complex event can thus be formally defined in terms of its components and the spatio-temporal relations that exist across these components. As shown in this example, the decomposition can be hierarchical and the elementary concepts and events may belong to different media forms. The "Anonymous" nodes in the event graph represent intermediate events that are not labeled otherwise. We have seen the mechanisms to define elementary concepts in the previous sections. In this section, we concentrate on the spatio-temporal relations that bind these concepts into complex event classes.

Incidentally, we note that the MPEG-7 Multimedia Description Scheme provides a mechanism to describe such complex events in a media instance using some spatio-temporal relations [166]. The relations, such as *left_of*, and so on, defined in MPEG-7, are, however, normative and are open to interpretation by the system designer. MPEG-7 ontologies [101] tend to formalize structural description of multimedia segments but continue to use the informal relations of MPEG-7 that cannot be used to create *formal* definition of event classes in an ontology.

MOWL adopts an approach proposed in [214] in creating formal definitions for complex events. The approach is an extension of Allen's interval algebra [5] and the representation scheme proposed in [153] and is explained in the following subsections.

5.5.1 Allen's Interval Algebra

Allen [4] introduced an interval algebra to describe the temporal relations between two events in terms of formal equality and inequality relations between their start and end times. A formal representation scheme for the relations was introduced in [153], where the relations are encoded as a 5-tuple of binary numbers. Each of the numbers represents the truth value of intersection of a secondary event duration with five *regions of interest* defined with respect to a reference (or primary) event as shown below:

$(-\infty, a), [a, a], (a, b), [b, b]$ and $(b, +\infty)$ where a and b represent the start

and end times of the primary event. For example, the interval $[c, d]$ in figure 5.6, which is confined to the region $(-\infty, a)$ with respect to the primary event $[a, b]$, is represented by the 5-tuple 10000 and is equivalent to the Allen's relation "$[c, d]$ *precedes* $[a, b]$." There can be $2^5 = 32$ binary representations as per Papadias's scheme. Assuming that both the events are finite, continuous, and not infinitesimally small, some of the representations are practically impossible, leaving exactly thirteen valid bit-patterns corresponding to the Allen's relations.

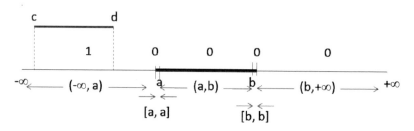

FIGURE 5.6: Binary encoding of interval relations.

Though these relations were originally defined for time, they can be readily extended to other dimensions, such as the space coordinates. Thus, two multimedia events can be related in 4D space (x, y, z, t) using a quadruplet $\langle R_x, R_y, R_z, R_t \rangle$ where each member of the quadruplet is a 5-tuple and represents a relation between the projections of the events in the corresponding dimension. This representation suffers from two limitations, the remedies for which, together with their solutions, are presented in the following subsections.

5.5.2 Concave Intervals

Although the specification shown above works well for convex intervals, it cannot handle the intersection of two events when one or both of the event spaces are concave.

In order to overcome this difficulty, a set of *containment relations* between two events A and B have been defined in [214]. There are four distinct possibilities, represented by the four containment propositions as follows:

$p := A \setminus (A \cap B) \neq \phi$: A is not completely included in B,

$q := B \setminus (A \cap B) \neq \phi$: B is not completely included in A,

$r := A \cap B \neq \phi$: A and B have some intersection, and

$s := A \cap^* B \neq \phi$: A and B has some regular intersection.

where $\cap^*$ denotes the regularized intersection [1].

These containment conditions give rise to six relations R_c between a primary event B and a secondary event A, each of which can be represented as 4-tuples in terms of parameters p, q, r and s as

R_{1100} (*outside*), R_{1011} (*contains*), R_{0111} (*inside*),

R_{1111} (*overlaps*), R_{1110} (*touching*), and R_{0110} (*skirting*)

as illustrated in figure 5.7.

Thus, the spatio-temporal relations between two events (with convex or concave contours) can be represented without ambiguity, using a 5-tuple:

$$R = \langle R_x, R_y, R_z, R_t, R_c \rangle$$

where R_x, R_y, R_z and R_t denote Allen's relations for the projections of the event in x, y, z and t axes respectively, and R_c denotes the containment relation. It is interesting to note that in many situations, it may not be necessary to specify all elements of R to specify a relation. For example, the *Inside* relation for a soccer video (see figure 5.5) can be defined as a logical conjunction of R_x, R_y and R_c as:

$$\langle R_x : R_{00100}, R_y : R_{00100}, R_c : R_{1100} \rangle$$

which should hold good in some frames of the video.

---

[1] $A \cap^* B = \textbf{closure}(\textbf{interior}(A \cap B))$ [60]. Unlike the normal intersection, the operation $\texttt{interior}(A \cap B)$ leaves out all the boundary pixels and retains only the interior pixels of the operation $(A \cap B)$. The operation $\texttt{closure}(S)$ results in boundary pixels that envelop the set of pixels S such that each boundary pixel is near at least one interior pixel.

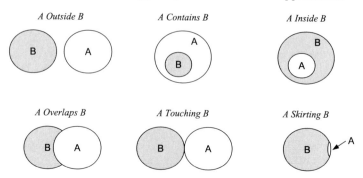

FIGURE 5.7: Distinct containment relations.

5.5.3   Accommodating Viewpoints

The relation so defined evaluates to crisp "*true*" or "*false*." Multimedia data processing, however requires some uncertain handling of such relations. For example, figure 5.8(a) provides a visual model of the `Tajmahal` as "dome *contained in* minarets," where the relation *contained in* is synonymous to "during" in Allen's terminology. Figure 5.8(b) and (c) provide two views of the monument, for which the specified condition is not an exact invariant. This variation needs to be accounted for, and these images should be recognized as instances of the monument during concept recognition.

The binary representations for the five intersections between the event intervals as proposed in [153] have been replaced by five fuzzy membership values for each of the dimensions x, y, z and t, to account for the uncertainties arising out of differences in viewpoints. Let us examine the representation in one of the dimensions, say x, in closer details. While we refer to figure 5.8 for illustration, the arguments are valid for any spatial or temporal dimensions.

In order to compute the fuzzy membership values, five functions, T, U, V, W, and X, each corresponding to one of the five regions of interest with respect to the prime event, have been defined. Each of the functions is defined in the domain $(-\infty, \infty)$ and is piecewise linear with respect to the start and the end points of the primary event. Each of them can assume values in the range $[0, 1]$. For example, a typical set of definitions for these functions for the X dimension can be as follows:

$$\text{T} ::= (-\infty, 1), (t_1, 1), (t_2, 0), (+\infty, 0)$$
$$\text{U} ::= (-\infty, 0), (u_1, 0), (u_2, 1), (u_3, 1), (u_4, 0), (+\infty, 0)$$
$$\text{V} ::= (-\infty, 0), (v_1, 0), (v_2, 1), (v_3, 1), (v_4, 0), (+\infty, 0)$$
$$\text{W} ::= (-\infty, 0), (w_1, 0), (w_2, 1), (w_3, 1), (w_4, 0), (+\infty, 0)$$
$$\text{X} ::= (-\infty, 0), (x_1, 0), (x_2, 1), (+\infty, 1)$$

where each of the tuples (x, v) represents a point x on the X axis and a value v attained at that point. Three of the functions, T, V and X, are depicted

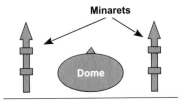

(a) Model of Tajmahal: dome *between* minarets

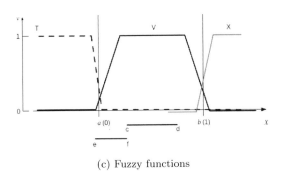

(b) Two views of Tajmahal from different perspectives (source: Wikipedia)

(c) Fuzzy functions

FIGURE 5.8: Fuzzy functions for defining inexact spatial relation.

in figure 5.8(d). The X axis is normalized with respect to the primary event $[a, b]$, that is, its start and end are represented by 0 and 1, respectively. The constants t_1, t_2, u_1, u_2 and so on represent some specific points on the axis, generally close to but not necessarily coinciding with the start and end points of the primary event. The value of the function is linearly interpolated between any two successive points, where they are specified.

The value of the five fuzzy membership functions that represent R_x and the spatial relation of a secondary event with respect to a primary one are represented by a tuple of five real variables as $\langle \mathfrak{t}, \mathfrak{u}, \mathfrak{v}, \mathfrak{w}, \mathfrak{x} \rangle$, each of which can assume a value in the range $[0, 1]$. The value of each variable is determined by the maximum value of the corresponding fuzzy membership function attained in the intersection interval between the secondary event and the corresponding region of interest. For example, consider the interval $[c, d]$ depicted in figure 5.8(d), which represents the position of the dome with respect to the minaret pair in figure 5.8(b). According to Allen's algebra, the interval $[c, d]$ is *contained in* $[a, b]$ and can be represented by the 5-tuple 00100 according to the representation scheme [153]. In the closed intersection interval $[[c, d] \cap [a, b]]$,[2] V attains a maximum value of 1. For all other functions, T,U,W and X, the maximum value attained is 0. Thus, the fuzzy representation of the relation becomes $(0, x, 1, x, 0)$,[3] which is in consonance with the crisp representation such as 00100. However, we can see the difference when we consider the interval $[e, f]$, which represents the position of the dome with respect to the minarets as depicted in figure 5.8(c) (somewhat exaggerated for the purpose of illustration). According to Allen's relations, $[e, f]$ is overlapped by $[a, b]$, and thus does not satisfy the model of Tajmahal as presented in figure 5.8(a). However, the representation of the relation with fuzzy membership functions is $(0.4, x, 0.9, x, 0)$. Thus, if the relation "between" is defined in terms of these three functions (ignoring the other two) as 0x1x0, the interval $[e, f]$ has a distance 0.1 (according to fuzzy algebra) from the definition and can still satisfy the model, though with a lower confidence value. Fuzzy conjunction rules can be used to combine membership functions in multiple dimensions.

The regions of interest for containment functions need a slightly different definition. The normalized regions of interest for the containment relations can be represented as follows: $\frac{|A \backslash A \cap B)|}{|A|}$, $\frac{|B \backslash A \cap B)|}{|B|}$, $\frac{|A \cap B|}{min(|A|,|B|)}$ and $\frac{|A \cap^* B|}{min(|A|,|B|)}$.

Accordingly, four fuzzy membership functions $\mathcal{P}$, $\mathcal{Q}$, $\mathcal{R}$ and $\mathcal{S}$ are required for the containment relations. As in the case of spatial and temporal dimensions, the membership functions are defined as piecewise linear functions, and the membership value of any event for any of the six containment relations with respect to a primary event can be computed using fuzzy intersection rules.

5.5.4 Discussions

Complex event classes can be defined in MOWL using a hierarchical composition with elementary concepts and spatio-temporal relations that bind them. MOWL adopts the approach described in [214] for formal definition of spatio-temporal relations. The approach is an extension of the representation

---

[2]Recall that V corresponds to the region of interest $[a, b]$.
[3]x represents unknown or "don't care."

scheme proposed in [153] for Allen's interval algebra [4]. The extensions relate to

- Definition of a containment function that is necessary to distinguish events with concave contours in multidimensional space, and

- Introduction of fuzzy membership values instead of binary values for a secondary event to belong to a region of interest of a primary event.

Following this approach, a spatio-temporal relation can be defined in MOWL with a 5-tuple $\langle R_x, R_y, R_z, R_t, R_c \rangle$ in the most general case. Each of R_x, R_y, R_z and R_t represents a relation in spatial (x, y, z) and temporal (t) dimensions and is represented by a 5-tuple of real numbers $\langle \mathfrak{t}, \mathfrak{u}, \mathfrak{v}, \mathfrak{w}, \mathfrak{x} \rangle$. Each of the numbers represents a fuzzy membership value for a secondary event to belong to one of the five regions of interest of a reference (primary) event. R_c represents the containment relation and is represented by a 4-tuple $\langle \mathfrak{p}, \mathfrak{q}, \mathfrak{r}, \mathfrak{s} \rangle$.

The event definition in MOWL is more general in nature than Allen's relation and offers flexibility to define multimedia event classes. Note that the semantics of any relation can be changed by changing the members of the 5-tuple or by changing the fuzzy functions. Also, note that the crisp Allen's relations are special cases of this generic representations and can be derived by defining the fuzzy membership functions as appropriate step and impulse functions.

5.6 MOWL Language Constructs

The language constructs of MOWL are based on RDF/RDFS. A few new datatypes are also defined that have specific contextual semantics. These language constructs are described with examples in the next few sections.

5.6.1 Concepts and Media Properties

MOWL declares three named classes to represent **concepts, media patterns** and **media examples**, and these become the superclasses for all concepts and media observables defined in a MOWL ontology. The MOWL class <mowl:Concept> is used to describe the concepts that represent real-world entities and their hierarchical structure. Classes <mowl:MediaPattern> and <mowl:MediaExample> are for the media patterns and media instances, which as mentioned earlier are the manifestation of concepts in media form. **Media-based relations** that relate concepts to media patterns and media examples are <mowl:hasMediaPattern> and <mowl:hasMediaExample>. <mowl:hasMediaPattern> is a causal relation, so MOWL allows for probabilities to be attached to the entities in this relation. The schema definition of all MOWL constructs is given in appendix A.

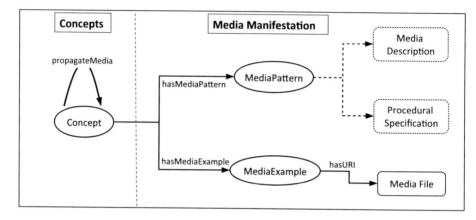

FIGURE 5.9: Concept and media observables as defined in MOWL.

A media pattern can be specified through a standard descriptor or through a procedural specification. The latter can be part of the media pattern description or be located through a URI specified in the description. Similarly a media example associated with a concept can be specified through a URI pointing to the location of the media file or through a procedural specification of the example detection module. <mowl:hasURI> property attaches a URI that points to a media file or to a procedural specification.

5.6.2   Media Property Propagation

Relations between the concepts play an important role in concept recognition. For example, an important cue to the recognition of a medieval monument can be the visual properties of the stone it is built with. As another example, a classical dance form is generally accompanied by a specific form of music. Thus, detection of media properties characterizing the music form is an important cue to recognition of the dance form. In order to enable such reasoning, MOWL allows definition of a class of relations that imply "propagation" of media properties. These relations are different from hierarchy relations, which also permit the flow of media properties by virtue of their semantics. To specify such relations that do not imply a concept hierarchy but allow propagation of media properties and examples across connected concepts in the ontology, MOWL defines a mowl:propagateMedia property.

Figure 5.10 shows an example of the media propagation that happens due to the <mowl:propagateMedia> relation between two concepts in a MOWL ontology. The figure shows an RDF graph of the ontology concepts and also the corresponding MOWL snippet. The format for ontology visualization is such that the concepts are shown as ellipses and the media properties as rectangles.

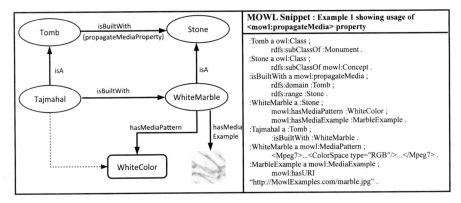

FIGURE 5.10: Media properties of concept `Marble` get associated with Tajmahal.

We have used the Terse RDF Triple language (Turtle) [13] due to its compact and natural text form to write the MOWL snippet. This convention is followed through rest of the book. In figure 5.10, an individual relation *isBuiltWith* is defined as an instance of `<mowl:propagateMedia>` property, with the class `Tomb` as the domain and the class `Stone` as the range. The ontology also specifies that `Tajmahal`, an instance of `Tomb`, *isBuiltWith* `Marble`, which is a kind of `Stone`. Since *isBuiltWith* is a `<mowl:propagateMedia>` property, the media properties of `Marble` propagate to `Tajmahal`, and the domain concept `Tajmahal` can be recognized in a media document on the basis of evidential presence of white color of marble.

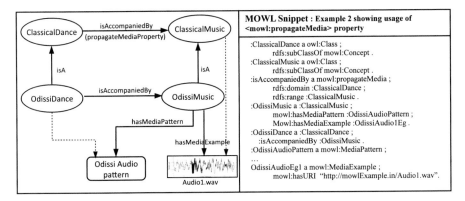

FIGURE 5.11: The media properties of `OdissiMusic` get associated with `OdissiDance`.

In the domain of Indian classical dance, a video of `OdissiDance` performance has high probability of being *accompanied by* a music score (recognizable as an audio pattern) from the `OdissiMusic` subclass of Indian classical music. This relation *isAccompaniedBy* can be defined as an instance of the `<mowl:propagateMedia>` property. Then the media properties of concept `OdissiMusic` which is the *audio pattern* recognizable as `OdissiAudioPattern`, *propagate* to the concept of `OdissiDance` (figure 5.11). Media examples propagate upwards in a hierarchical relation in MOWL, so as figure 5.11 shows, `Audio1.wav`, which is an audio example of `OdissiMusic`, is also an example of its parent class `ClassicalMusic`. Later we see in section 5.8.2 how this propagation helps in the recognition of concept `OdissiDance` in a video.

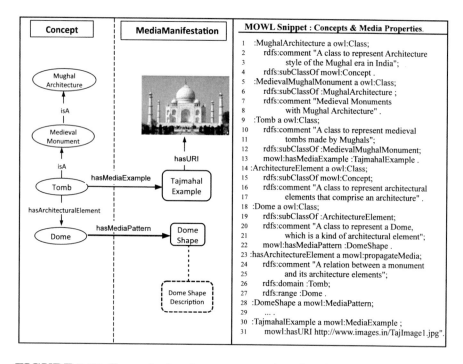

FIGURE 5.12: Example showing concept and media observables in a MOWL ontology.

5.6.3 Uncertainty Specification

Knowledge in several domains like medicine, art and heritage is characteristically uncertain. Such knowledge is impossible to represent in web ontology languages like RDF or OWL and reasoning with it cannot be done with crisp description logic. A Bayesian network is a directed acyclic graph, in which each node represents a variable that can assume any one of finitely many states. The uncertainty in a Bayesian network is represented by a set of conditional probability tables, $P(X_i \mid \pi_i)$ for every node X in the network. Here, X_i represents a state of the variable X and π_i a state of its parent variables. A Bayesian network has a causal interpretation: $P(X_i \mid \pi_i)$ implies that the state π_i of the parent set causes the state X_i of variable X with a probability P. In multimedia domain, the causal relations implied by a Bayesian network hold good, as a media property can be considered to be the manifestation of an abstract concept in a multimedia document. For example a *sunny day* manifests itself as a *blue* sky in a landscape photograph, or an *abnormal growth* in the brain manifests in a certain shape observed in an MRI image. Each of these manifestations is, however, associated with some probability. A photograph taken at dawn on a sunny day may not depict a blue sky. Similarly, a particular shape in an MRI image may not be an irrefutable evidence for an abnormal growth. These uncertainties can be represented by a conditional probability table between boolean variable pairs $P(M \mid C)$ and $P(M \mid \neg C)$, where M denotes a media pattern and C denotes a concept. Similar uncertain causal relations exist between concepts in the multimedia domain — for example, the sound of a *cheer* will materialize in a football video after a *goal scored* only with a certain probability. The uncertainty in a domain can be characterized as a joint probability distribution over the variables corresponding to the domain classes and individuals. A joint probability distribution can be factorized into several component factors, and each factor is a distribution involving fewer variables. The factorization makes it possible to specify the uncertainty of a variable as a conditional probability distribution over a set of variables that are its direct causes (its parent variables in the ontology graph).

MOWL provides for specification of uncertainty in a multimedia domain by providing special constructs for defining conditional probability tables (CPTs) and associating them with relations. The MOWL constructs that support uncertainty specification are described in the following schemas:

- <mowl:hasCPT> property associates a CPT defined by <mowl:CPTable> with a MOWL concept or media pattern.

- <mowl:conditionedOn> property specifies the parent or parents of a concept, conditioned on which the CPT entries are based.

- <mowl:CPTable> class allows the definition of a conditional probability table in the ontology. A CPT must mention the cause(s) or the parent(s) of the concept with which it is associated. A CPT has many rows, each of which has the probability values for the two states of the concept based on a particular combination of the states of its parents.

- <mowl:hasRow> property that specifies that the CPTable has CPRows.

- <mowl:CPRow> is the class that represents a row of the CPT. It has two data-type properties

- <mowl:parentStates>, which specifies the combination of the states of the parents of the concept. Mostly there is one parent, and the state is either 1 or 0, meaning the parent concept is present or absent.

- <mowl:probValues>, which is a vector of two probability values $P(X_i = 1 \mid \pi_i)$ and $P(X_i = 0 \mid \pi_i)$, between 0 and 1, both inclusive.

A point to be noted while specifying the conditional probabilities is that a concept or media node in a MOWL ontology can have only two states. Thus it can have values true or false, absent or present, or 0/1. This implies that the number of entries in the probability vector in each row of the CPT is always two, but the number of rows in the CPT depends upon the number of parents of the node. In case a concept node has multiple parents, and its conditional probabilities pertaining to one parent are independent of those conditioned on the other parent, then the concept can have multiple CPTs, with one CPT per parent. In another case, when a concept node has multiple parents but they are dependent on each other, then the concept can have a single CPT with multiple rows depending on the combination of states of the parents. For 2 states of each parent, there will be 2^n rows for n parents. For example, a CPT for a concept with one parent will have two rows, one for each state of the parent; CPT with two interdependent parents will have four rows, and so on.

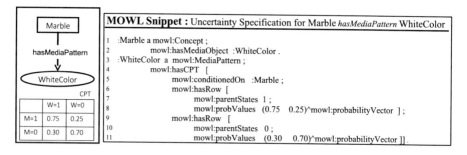

FIGURE 5.13: MOWL constructs for the uncertainty specification.

To illustrate uncertainty specification using MOWL, consider the MOWL snippet in figure 5.13. It defines the CPT for the media pattern `WhiteColor` conditioned on the concept `Marble`. This CPT encodes the uncertainty related with the observation that marble generally has white color but not all types of marble have a white color.

The proposed MOWL specification of uncertainty overcomes the limitation of [48] in which the uncertainty of a variable could be conditioned on only a single variable. It is not mandatory to specify the conditional probability tables or the prior probability with every concept in MOWL. In the absence of these specifications, a crisp relation between the nodes can be assumed and appropriate conditional probabilities can be assigned. This is justified by the observation that inferencing in Bayesian network is more sensitive to network topology than the conditional probability tables [159]. Thus, the quality of inferencing is unlikely to have significant degradation with such default probabilities. The default prior probabilities of a node are assigned the unbiased value of $\frac{1}{2}$.

5.6.4 Spatio-Temporal Relations

MOWL defines a subclass <mowl:ComplexConcept> of <mowl:Concept>, which represents a concept that is composed of exactly two component concepts related through a spatial or a temporal relation. Every complex concept is defined by a spatio-temporal relation or *predicate* and two concepts — one the *subject* of the predicate relation and the other the *object* of the predicate. The three properties of the <mowl:ComplexConcept> class are:

- mowl:Subject
- mowl:hasPredicate
- mowl:Object

For example, a soccer goal can be represented as a complex concept with *subject* "ball," *predicate* "inside" and *object* "goalpost." As this class extends the class of concepts in MOWL, the concepts involved in the spatio-temporal relation can themselves be complex concepts. This helps in defining complex concepts or events like Goalscore in soccer where more than one spatial and temporal relation are involved. In general, a complex concept can have other complex concepts as its constituents, either subject or object. As a result, arbitrarily complex concepts can be built in a hierarchical fashion. Figure 5.14 shows such a graphical representation.

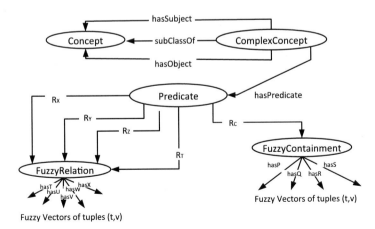

FIGURE 5.14: Graphical representation of a complex concept that represents spatio-temporal events in MOWL.

MOWL defines a named class <mowl:Predicate> to represent spatio-temporal predicates. Spatio-temporal relations like inside and followedBy can be defined as instances of this class. These predicates, being relations, should ideally be defined as a property in an ontology language, say a sub-property of <rdf:Property> or <owl:ObjectProperty> with a domain and range. From section 5.5, recall that the spatio-temporal relations in MOWL are represented by a 5-tuple $\langle R_x, R_y, R_z, R_t, R_c \rangle$, each of which is represented with a tuple of fuzzy functions. These relations need to be reified and described with the help of their own properties. As such reification is not possible in RDF representation, these predicates have been defined as classes instead of a property in MOWL.

MOWL defines two classes to formally define the spatio-temporal relations in terms of generalized Allen's relations and containment relations:

- <mowl:FuzzyAllen>: Its subclasses include all possible Allen's relations like *precedes, finishedBy, overlaps,*, etc. (equivalently, represented as bit strings like *R11000*). The schema definition for this class is shown in appendix A.

- <mowl:FuzzyContainment>: Its subclasses include all fuzzy containment relations like *outside, contains,* etc. (equivalently represented as bit strings like *R1011.*)

It is also possible to define new relations with user-defined semantics as subclasses of these two classes. The semantics of such relations are governed by the definition and use of associated fuzzy membership functions (see the example in section 5.5). Each of the fuzzy membership functions (T, U, V, W and X) associated with the five regions of interest for each of the space and time axes (x, y, z and t) is a vector of tuples (t, v), where v is the fuzzy membership value at the point t on the axis. While t can take any arbitrary values in $(-\infty, +\infty)$, v can assume values in the closed range $[0, 1]$. The fuzzy functions for the containment relations are similarly defined.

MOWL defines some special XML Schema data types to define these fuzzy functions and the values:

- <xsd:probValType> data type defines a decimal value between 0 and 1, both inclusive, for v value of the tuple (t, v).

- <xsd:axisPtType> defines a data type that can be a string with values "-infinity" or "+infinity" or a decimal for t value of the tuple (t, v).

- <xsd:tupleType> defines a data type to represent a tuple (t, v).

- <xsd:FuzzyVector> defines a vector of tuples.

The properties for attaching these fuzzy-valued functions to Allen's and containment relations are defined in the schema in appendix A, as a part of the MOWL schema definition. It may be noted that there may not be a unique way

MOWL Snippet : *GoalScore* event defined in 2 ways.	MOWL Snippet : *inside* predicate definition.

```
1  :Ball  a  owl:Class ;
2       rdfs:subClassOf  mowl:Concept ;
3       ...
4  :GoalPost  a  owl:Class ;
5       rdfs:subClassOf  mowl:Concept ;
6       ...
7  :Cheer  a  owl:Class ;
8       rdfs:subClassOf  mowl:Concept ;
9       ...
10 :inside  a  mowl:Predicate ;
11      ...
12 :followedBy  a mowl:Predicate ;
13      ...
14 :Ball_in_Goal  a  mowl:ComplexConcept ;
15      mowl:Subject    :Ball ;
16      mowl:Predicate  :inside ;
17      mowl:Object     :GoalPost .
18 :GoalScore  a  mowl:ComplexConcept ;
19      mowl:Subject    :Ball_in_Goal ;
20      mowl:Predicate :followedBy ;
21      mowl:Object     :Cheer .
```

```
14 :GoalScore  a  mowl:ComplexConcept ;
15      mowl:Subject  [
16          mowl:Subject    :Ball ;
17          mowl:Predicate  :inside ;
18          mowl:Object     :GoalPost . ] ;
19      mowl:Predicate :followedBy ;
20      mowl:Object     :Cheer .
```

```
1  :inside  a  mowl:Predicate ;
2       mowl:RX  :R00100 ;
3       mowl:RY  :R00100 ;
4       mowl:RC  :contains ;
5  :R00100  a  mowl:FuzzyAllen ;
6       mowl:has_T [
7           axisPt "-infinity" ; value "1" ;
8           axisPt "-0.15" ; value "1" ;
9           axisPt "0.10" ; value "0" ;
10          axisPt "+infinity" ; value "0". ] ;
11      mowl:has_U [
12          axisPt "-infinity" ; value "0" ;
13          axisPt "-0.25" ; value "0" ;
14          axisPt "-0.10" ; value "1" ;
15          axisPt "0.10" ; value "1" ;
16          axisPt "-0.25" ; value "0" ;
17          axisPt "+infinity" ; value "0" . ] ;
18      mowl:has_V [ ... ] ;
19      mowl:has_W [ ... ] ;
20      mowl:has_X [ ... ] .
21 :contains  a  mowl:FuzzyContainment ;
22      mowl:has_P [
23          axisPt "-infinity" ; value "0" ;
24          axisPt "0.8"     ; value "1" ;
25          axisPt "+infinity" ; value "1" .] ;
26      mowl:has_Q [ ... ] ;
27      mowl:has_R [ ... ] ;
28      mowl:has_S [ ... ] .
```

FIGURE 5.15: MOWL snippets for `GoalScore` event in a soccer ontology.

to define an event class using spatio-temporal descriptors of MOWL. As an example, the first MOWL snippet in figure 5.15 shows two different ways of representing a `GoalScore` event using MOWL spatio-temporal constructs. In the first case (in the top unshaded part), the event `Ball-In-Goal` is first defined as a complex concept with subject `Ball`, predicate `inside` and object `GoalPost`. Then the spatio-temporal event `GoalScore` is defined with two component concepts `Ball-In-Goal` and `cheer` related by the spatio-temporal relation `followedBy`. The semantics of spatio-temporal predicates `followedBy` and `inside` can be formally defined using MOWL spatio-temporal constructs. In this definition of `GoalScore`, the subject node `Ball-In-Goal` is defined explicitly and not just as an unnamed node as is done in the lower shaded part. This kind of explicit definition for a complex concept in MOWL allows its reuse as a component in other complex concept definitions as well. Another way to represent the `GoalScore` event in MOWL is as shown in the lower shaded portion of the first MOWL snippet. This is a more compact representation, as it shows how MOWL allows an unnamed event to be defined while defining complex concepts. Here the event `Ball_In_Goal` of the previous `GoalScore` definition is defined *inline* and is not named.

As an illustration of how MOWL allows new spatio-temporal relations to be defined, we show a MOWL snippet showing the definition for the spatio-temporal predicate `inside` from the `GoalScore` example in figure 5.15. Relation `inside` is shown with fuzzy functions for dimensions R_x, R_y, R_c. Similarly, the relation `followedBy` can be shown with fuzzy values for temporal dimension R_t. Figure 5.16 shows a graphical representation of these MOWL constructs.

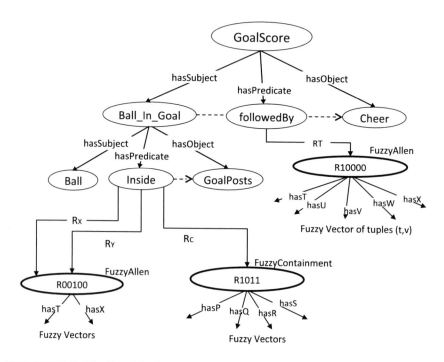

FIGURE 5.16: Graphical representation of the spatio-temporal constructs in MOWL for a `GoalScore` event in a soccer ontology.

5.7   The Observation Model

An observation model (OM) is constructed from a MOWL ontology. It represents a media-based description of a concept. It incorporates domain-specific contextual information and is used for concept recognition. In this section, we provide a formal definition of an OM and establish its semantics.

5.7.1   Defining an Observation Model

An observation model is a directed acyclic graph. It comprises several concept and media property nodes. The root node in the graph signifies the concept to be recognized in a multimedia document (also called the primary concept). The intermediate nodes refer to some related concepts, from which media properties may propagate to the root node (also called the secondary concepts). The leaf nodes represent observable media properties. Each node in an OM represents a binary variable that can assume the values "*true*" or "*false*." The edges in the graph represent causal relations. Each node of an observation model is associated with a conditional probability table (CPT) that represents conditional probability values for the states of the node given the states of its parent nodes. Another interpretation of the CPTs is that they represent the "causal strength" between the concepts and their individual media properties.

An OM comprises a small subset of concept and media property nodes present in the ontology. Thus, it defines the boundaries of a concept recognition problem. The acyclic property of the graph is necessary to prevent indefinite computational loops in the reasoning scheme.

An OM Ω_{C_p} for the primary concept C_p is defined as a graph

$$\Omega_{C_p} = \{\mathcal{V}, \mathcal{E}\}$$

where $\mathcal{V}$ is a set of vertices, and $\mathcal{E}$ is a set of edges.

The set of vertices $\mathcal{V}$ is composed of three subsets

$$\mathcal{V} = \{\mathcal{C}_p, \mathcal{C}_r, \mathcal{M}\}$$

where $\mathcal{C}_p$ is the primary concept for which the OM is constructed, $\mathcal{C}_r$ are the related concepts, and $\mathcal{M}$ is the set of expected media patterns. $\mathcal{C}_r$ includes the complex concepts, which are special concept nodes.

The expectation for observations of the primary concept $\mathcal{C}_p$ arises out of the observations for the concepts $\mathcal{C}_r$ as well, as these are related to $\mathcal{C}_p$ in the domain through the following relations:

$$\mathcal{C}_p \xrightarrow{\mathcal{R}_P} c, \quad \mathcal{C}_p \xrightarrow{\mathcal{R}_H} c$$

where $c \in \mathcal{C}_r$. $\mathcal{R}_P$ and $\mathcal{R}_H$ represent "property propagation" and "hierarchy" relations in the ontology that signify media property and example propagation.

Let $\mathcal{C}$ be the set of all concepts in the OM, that is, $\mathcal{C} = \mathcal{C}_p \cup \mathcal{C}_r$, then $\mathcal{M}$ is a union of the set of the media patterns associated with each of the concepts c in $\mathcal{C}$.

The set of directed edges $\mathcal{E}$ contains links from primary concept to the related concepts and from concepts to their associated media patterns. If $L_{A,B}$ denotes a link from a node A to a node B, the set of edges in the graph comprises the following three subsets:

- $\mathcal{L}_{PR} = \{L_{\mathcal{C}_p, c \in \mathcal{C}_r}\}$: The links connecting the primary concept to the related concepts.

- $\mathcal{L}_{PM} = \{L_{\mathcal{C}_p, m \in \mathcal{M}}\}$: The links connecting the primary concept to its media properties.

- $\mathcal{L}_{RM} = \{L_{c \in \mathcal{C}_r, m \in \mathcal{M}}\}$: The links connecting the secondary concepts to their respective media properties.

Thus the set of all directed edges can be expressed as an union of these three subsets:

$$\mathcal{E} = \mathcal{L}_{PR} \cup \mathcal{L}_{PM} \cup \mathcal{L}_{RM}$$

5.7.2   Semantics of MOWL Relations for Constructing OM

In this section, we discuss the semantics of the different relations in MOWL that are used for construction of an OM. An ontology is a graph and can be interpreted as a semantic network, with nodes corresponding to concepts, and the links corresponding to the relations between concepts. A MOWL ontology graph contains the following entities and relations:

1. A set of concepts $\mathcal{C}$, with a special subclass of complex concepts

2. A set of media patterns $\mathcal{M_P}$

3. A set of media examples $\mathcal{M_E}$

4. A set of media-based relations $\mathcal{R}_M: \mathcal{C} \xrightarrow{\mathcal{R}_{M_P}} \{\mathcal{M}_D, \mathcal{M_E}\}$
   where $\mathcal{R}_{M_P}$ are probabilistic causal relations, as a concept *causes* its media properties to be manifested in a multimedia document, but with some uncertainty

5. A set of "hierarchy" relations $\mathcal{R}_H$, which relate a subconcept to its superconcept. Subconcepts of a concept are the subclasses of the class representing the concept, as well as individuals belonging to it: $\mathcal{C} \xrightarrow{\mathcal{R}_H} \mathcal{C}$

6. A set of "propagate media" relations $\mathcal{R}_P$ that relate a concept with other concepts for propagation of media properties: $\mathcal{C} \xrightarrow{\mathcal{R}_P} \mathcal{C}$

There are other relations that may exist in the ontology, and that are neither hierarchical nor media propagate relations, and thus are not causal. For example Tajmahal *built_by* Shahjahan in the ontology shown in figure 5.3 is not a causal relation. These relations are not added to the OM as they do not contribute to the inferencing framework.

We discuss the semantics of the MOWL relations representing "hierarchy" ($\mathcal{R}_H$) and "media property propagation" ($\mathcal{R}_P$) that are important for construction of the OM. The part of the MOWL ontology graph considered for OM construction (Γ) can be seen as an overlay of two subgraphs — a hierarchy subgraph Γ_h of the hierarchy relations, and a propagation subgraph Γ_p of media propagate relations. [4]

---

[4] The acyclic properties of these subgraphs impose a certain ordering of concepts.

Let a relation $\mathcal{R}$ represent either of these two MOWL relations. If $\mathcal{C}_1 \xrightarrow{\mathcal{R}}$,$\mathcal{C}_2$ then we call $\mathcal{C}_1$ a *descendant* of $\mathcal{C}_2$ in its respective subgraph, while $\mathcal{C}_2$ is called an *ancestor* of $\mathcal{C}_1$. By the semantics of inheritance, $\mathcal{C}_1$ inherits all properties of $\mathcal{C}_2$; these include all the media properties of $\mathcal{C}_2$ as well. This is true for both the relations. It is important to note that when these edges get added to the OM, they lose the semantics of the relation that they represent between two concepts in the ontology. In the OM, a link between two concepts simply represents a causal relation in the Bayesian network. Further we discuss the semantics of each MOWL relation towards OM construction individually:

Hierarchical Relations $\mathcal{R}_H$:

These relations comprise the "subclass of" and "instance of" relations between two concept nodes in an ontology. For simplicity, we will denote these properties with a common name "isA" in the ontology. Figure 5.17 shows Tomb as a subclass of Monument and Tajmahal as an instance of Tomb. These can written as Tajmahal $\xrightarrow{\mathcal{R}_H}$ Tomb, and Tomb $\xrightarrow{\mathcal{R}_H}$ Monument. Note that in the hierarchy, Tomb is actually a descendant of Monument, but the direction of the link representing $\mathcal{R}_H$ relation in the graph is from Tomb to Monument. The link between an instance and its class is also *from* the instance (Tajmahal) *to* its class (Tomb).

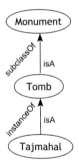

FIGURE 5.17: Hierarchical relations isA.

This direction of the hierarchical relation between the two concepts is maintained when the relation is added as a link to the OM, since it correctly represents the direction of the causal link. That is, Tajmahal $\rightarrow$ Tomb in the OM implies that observation of Tajmahal *causes* observation of Tomb, which is a correct inference.

Media properties are inherited by the descendant, but media examples can be propagated to the ancestor in a hierarchy relation. For example, as shown in figure 5.12, concept Tomb has a structural component Dome associated with it as a media property. Tajmahal as a subconcept of Tomb inherits this property, and is expected to have a Dome in its structure. On the other hand, media examples such as images of Tajmahal are also the media examples of its ancestor Tomb. But the reverse of these rules is not true : media properties of a descendant *do not propagate* to the ancestor, and media examples of an ancestor *are not inherited* by the descendant. This is illustrated by the observations that photographs (media examples) of Tomb may not all be of *Tajmahal*, and so cannot be associated with the latter. Similarly the media property *color=white* (refer to figure 5.10) of Tajmahal does not propagate to its ancestor Tomb, as other tombs may be made of different colored material.

A point to be noted here is that a descendant in hierarchy, whether a subclass or instance of the concept, may specialize some media properties of the latter. For example Tajmahal has structural component "Dome between Minarets", that contains the structure Dome but in a spatial relation with other structural components (minarets). If this is the case, then Tajmahal need not inherit the media property Dome from its ancestor Tomb. This fact has to be considered in construction of the OM.

Propagate Media property relations $\mathcal{R}_P$:

These relations refer to instances of MOWL property <mowl:propagateMedia> between two concepts in a MOWL ontology. This property is defined for those relations that allow media properties to propagate from the range of the property to the domain. In the propagation subgraph, if $C_1 \xrightarrow{\mathcal{R}_P} C_2$, then media properties of ancestor C_2 propagate to the descendant C_1. For example, as shown in figure 5.4, Tajmahal $\xrightarrow{\mathcal{R}_P}$ Marble is true and so media property "color=white" of ancestor Marble propagates to descendant Tajmahal. However, media examples do not propagate across $\mathcal{R}_P$ relations; for example, media examples of Marble are not examples of Tajmahal.

Media Relations $\mathcal{R}_{M_O}$:

Media properties are media manifestations of concepts in MOWL. The manifestation is in the form of an observable media pattern or media examples. Media patterns and examples can be defined in terms of procedural specifications of media detection tools to detect the patterns or to classify the examples while defining the ontology. These tools return a belief value for the observation of the media pattern in a multimedia document or a similarity score for its comparison with the classified models. The leafnodes in the OM constitute the set of expected media patterns $\mathcal{M}$ for the concepts in the OM. All the media properties of the primary concept are added to its OM recursively.

The set of related concepts of the primary concept c in the OM can be defined as $\mathcal{C}_r = \{\alpha_h(c), \beta_h(c), \beta_p(c)\}$, where $\alpha_h(c)$ represents the set of descendants and $\beta_h(c)$ the set of ancestors of c in the hierarchy subgraph, wherein $\beta_p(c)$ represents the set of ancestors of c in the propagation subgraph. For related concepts, the following rules apply:

- For $\alpha_h(c)$: Their media examples only are added to the OM.

- For $\beta_h(c)$ and $\beta_p(c)$: Their media patterns only are added to the OM. Media examples of these related concepts are not added to the OM.

Spatio-temporal Relations and Complex Concepts:
Spatio-temporal relations bind two or more elementary concepts in the form of complex concepts. Each complex concept is added as single node in the OM. These nodes partition an observation model in mutually exclusive Bayesian networks. The reasoning within these nodes follows a different pattern than that of the Bayesian network. The reasoning scheme for concept recognition is detailed in section 5.8.2.

5.7.3   Computing CPTs in the OM

The CPTs in the OM are provided by instances of the <mowl:CPTable> construct encoded in the MOWL ontology based on domain knowledge with respect to a model of the world. The OM has the primary concept as its root node, and the related concepts are added recursively to it as its descendants to form a graph structure. We refer to that part of the ontology graph, which is considered for construction of the OM, as Γ. For each pair of adjacent nodes A and B in the OM, where causal relation $A \rightarrow B$ implies that A is the parent of B (that is A causes B), CPTs are computed as follows:

- If B is a child of A in Γ as well as in the OM, the conditional probabilities $P(B \mid A)$ and $P(B \mid \neg A)$ are obtained directly from Γ.

- If A is a child of B in Γ, but its parent in the OM, the conditional probabilities $P(B \mid A)$ and $P(B \mid \neg A)$ are computed as posterior probability values of A when B is instantiated to *true* and *false*. Assuming an *apriori* probability of $\frac{1}{2}$ for A, we apply Bayes' theorem for the computation. If $\beta = P(B \mid A)$ and $\rho = P(B \mid \neg A)$ in Γ, then CPT values in the OM are given by

$$\beta' = \frac{\beta}{\beta + \rho}, \quad \rho' = \frac{1 - \beta}{1 - \rho - \beta}$$

- Multiple connectivity may exist in Γ; that is, a concept node may have multiple parents. In such a case, the node is replicated in the OM (and not multiply linked) since the conditional probabilities of the individual links are available (rather than joint probability distributions). Thus the constructed OM is a tree.

5.8   MOWL Inferencing Framework

An ontology representation scheme is always complemented with an inferencing framework that specifies how the ontology can be used to make new inferences with the knowledge encoded in the ontology. It may be noted that the specified inferencing framework may not be the only way to exploit the domain model represented by the ontology. The inferencing scheme in MOWL presented in this section has been motivated by the need to construct observation models for concepts and to recognize concepts by observing expected media patterns in the multimedia documents. Thus the inferencing framework is driven by the observations of concept-specific media patterns. The semantics of the additional language constructs described in the previous section play an important role in the inferencing framework. The knowledge available in a MOWL ontology is used to construct an observation model for a concept, which is in turn used for concept recognition. This requires two stages of reasoning:

Reasoning for derivation of observation model for a concept:
This requires exploring the neighborhood of a concept and collating media properties of neighboring concepts, wherever media property propagation is implied. The resultant observation model of a concept is organized as a Bayesian network (BN). The BN so created reflects the causal relation between the concept and the media properties that can be expected in a manifestation of the concept in a multimedia document. The root node in the network represents the concept and the leaf nodes represent the observable media patterns. The intermediate nodes represent the related concepts. The probability distribution in the BN is derived from the joint probability distributions for the media properties available in the ontology.

Reasoning for concept recognition:
Once an observation model for a concept is created, it can be used for concept recognition. We use an abductive reasoning scheme that exploits the causal relations captured in the observation model. The Bayesian network is initialized with some *apriori* probability at the root node in the context of each document to be evaluated, based on the knowledge of the document collection. Then the pattern detectors for the patterns specified in the leaf nodes of the observation model are run on the multimedia document. When any of the media patterns are detected, the corresponding leaf nodes are instantiated and there is belief propagation in the Bayesian network. The concept is believed to have materialized in a document if the posterior probability of the root node exceeds a certain threshold as a result of detection of several patterns

and consequent instantiations of the leaf nodes. The posterior belief values for the documents can also be used for contextual ranking of the documents.

5.8.1   Constructing the Observation Model

Algorithms 1 and 2 detail the OM construction. The algorithm is based on the semantics of MOWL relations as discussed in section 5.7.2. The inputs to the OM construction algorithm are the ontology graphs, Γ, Γ_h and Γ_p, and the primary concept $\mathcal{C}_p$. The output expected is a Bayesian network — the observation model for $\mathcal{C}_p$.

The algorithm starts with an empty Bayesian network, then adds the primary concept $\mathcal{C}_p$ as its root node. A recursive procedure ADDMEDIAPROPERTIES is called to add all the media properties of $\mathcal{C}_p$. Then another recursive procedure ADDRELATEDCONCEPTS() is called to add the related concepts to $\mathcal{C}_p$. Recursion causes related concepts of related concepts to be added as well, till a leaf node is reached every time. Media properties of each related concept are added by calling the procedure ADDMEDIAPROPERTIES for each related concept. If a related concept or media property already exists in the network, a new replicated node for it is added. In order to avoid cycles in the OM, we check that the range concept of the link is not an ancestor of the domain concept before adding a new link,

Now steps are required to clean up the OM and remove those nodes that will not contribute to inferencing. Those concepts in the OM that do not have any media manifestation — that is, there is no media property attached to them — are leaf nodes in the OM. As there are no media properties available to be processed for these nodes, they are removed from the OM. In case there is any media pattern that has been *specialized* by another media pattern attached to the same concept, then the first media pattern is removed from the OM. This could happen when a concept has a media manifestation, but one of its descendants has a more generalized form of the same manifestation. This point is illustrated later with the example from ICD domain. Both these steps are done repeatedly till there are no such nodes left in the OM. After this, the CPTs are generated from the specifications in the ontology graph Γ, according to the rules discussed in section 5.7.3, and the OM is ready for the next step of reasoning with it.

5.8.2   Concept Recognition Using the OM

The structure of a typical OM is shown in figure 5.18. The root node (labeled "c" in the diagram) represents the primary concept, the ellipses represent secondary concepts and the rectangles represent media patterns. The shaded area in the figure represents a complex concept, that is, a spatio-temporal composition of elementary concepts. The existence of the complex node divides the graph into three partitions, each representing a Bayesian network. Inferencing in each of the Bayesian networks is by virtue of belief

---

Algorithm 1 Constructing an observation model for concept C_p. Part 1.

Inputs: a) MOWL Ontology graph Γ with Hierarchy subgraph Γ_h & Propagation sub-graph Γ_p

　　　　　　b) Primary Concept C_p

Output: Observation Model Ω_{C_p}

　1: **procedure** MAIN
　2:　　　Create an empty Bayesian network Ω_{C_p}
　3:　　　Root Node of $\Omega_{C_p} \leftarrow C_p$
　4:　　　AddMediaProperties(C_p)
　5:　　　AddRelatedConcepts(C_p)
　6:　　　**repeat**　　　　　　　　　　　　　　　　　　　▷ *Cleaning up the OM*
　7:　　　　　Remove any leaf node which is a concept
　8:　　　　　Remove any media pattern node which has been specialized for the same concept
　9:　　　**until** Ω_{C_p} has no concept leaf nodes and no specialized media patterns
10:　　　Compute CPTs from Γ and attach to nodes in Ω_{C_p}
11: **end procedure**
12: **procedure** ADDRELATEDCONCEPTS(C)
13:　　　**if** C is a leaf node in Γ **then**
14:　　　　　RETURN
15:　　　**end if**
16:　　　Compute $C_r = \{$ c : c is a related concept of C $\}$
17:　　　**for** $i = 1$ to $|C_r|$ **do**
18:　　　　　**if** c_i exists in Ω_{C_p} **then**
19:　　　　　　　**if** c_i is NOT an ancestor of C in Ω_{C_p} **then**　　　▷ *Avoiding cycles*
20:　　　　　　　　　Create a node c_j which duplicates node c_i
21:　　　　　　　　　Add c_j as a new child of C
22:　　　　　　　　　AddMediaProperties(c_j)
23:　　　　　　　　　AddRelatedConcepts(c_j)
24:　　　　　　　**end if**
25:　　　　　**else**
26:　　　　　　　Add c_i as a new child of C
27:　　　　　　　AddMediaProperties(c_i, C)
28:　　　　　　　AddRelatedConcepts(c_i)
29:　　　　　**end if**
30:　　　**end for**
31: **end procedure**　　　　　　　　　　　　　　　　　　▷ *Continued in part 2 ...*

---

propagation [141] from the observed nodes (the leaf nodes) to the root nodes of the respective network, amounting to abductive reasoning. Inferencing for a complex concept is governed by a fuzzy reasoning framework as discussed in section 5.5.

Belief Propagation in a Bayesian network:

Let us pretend for a while that the complex concepts do not exist and the entire OM is a Bayesian network. This will be true for each partition of the OM

Algorithm 2 OM construction algorithm. Part 2.

32: **procedure** ADDMEDIAPROPERTIES($\mathcal{C}$, $\mathcal{C}_{\vee}$)
33: **if** $\mathcal{C}$ is a leaf node in Γ **then** RETURN
34: **end if**
35: Compute $\mathcal{M}$ = { m : m is a media property of $\mathcal{C}$ }
36: **for** $i = 1$ to $|\mathcal{M}|$ **do**
37:   $Flag_{Add}$= FALSE
38:   **if** $\mathcal{C}$ is a descendant of $\mathcal{C}_p$ in Γ_h & m_i is a media example **then**
39:    $Flag_{Add}$ = TRUE
40:   **else if** ($\mathcal{C}$ is an ancestor of $\mathcal{C}_p$ in Γ_h or $\mathcal{C}$) & m_i is a media pattern **then**
41:    $Flag_{Add}$ = TRUE
42:   **end if**
43:   **if** $Flag_{Add}$ **then**
44:    Add m_i as a new child of $\mathcal{C}$
45:   **end if**
46: **end for**
47: **end procedure**

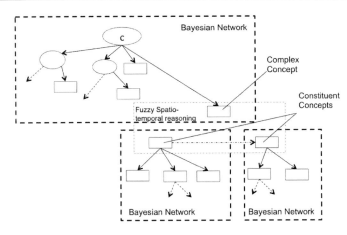

FIGURE 5.18: Observation Model - a Bayesian network.

created by the complex concepts, as illustrated in figure 5.18. Each concept node in the OM is considered a propositional node with two states: *"true"* (concept exists) and *"false"* (concept does not exist). The nodes signifying media properties are also binary. The truth value of a media property node signifies that the pattern has been detected.

For inferencing, the Bayesian network is initialized with some *apriori* probability at the root node in the context of each document to be evaluated. As the next step, the pattern detection tools (for the patterns specified in the leaf nodes of the observation model) are run on the document. When any of the media patterns are detected, the corresponding leaf nodes are instantiated.

The instantiation of some of the leaf nodes leads to belief propagation in the Bayesian network. The primary concept is believed to have materialized in a document if the posterior probability of the root node exceeds a certain threshold as a result of detection of several patterns and consequent instantiations of the leaf nodes.

Reasoning with Complex Concepts:

A complex concept in MOWL is a hierarchical composition, where each level of the hierarchy comprises two concepts interconnected with a spatial or temporal relation. In general, each of the constituent concepts can either be an elementary concept or a complex concept. The component concepts of the lowest level of the hierarchy are always elementary concepts. A reader may refer to an example composition of a "GoalScore" event presented in figure 5.5 for illustration.

Recognition of complex concepts happens in a bottom-up manner. The complex concepts at the lowest level of hierarchy are first examined. The constituent elementary concepts have their own respective OMs in the form of Bayesian networks, which are used for their recognitions. If both "subject" and "object" concepts are believed to have materialized, the spatial or the temporal relation between the concepts is explored. If the relation is found to exceed a threshold fuzzy membership value, the complex concept is believed to have materialized (see section 5.5). A complex concept so recognized serves as a building block for the next higher layer. This algorithm is applied recursively upwards, till the belief in the top-most item in the complex concept is recognized. Note that if the belief in any intermediate concept node is negated, the computations can be abandoned with a negative belief value for the top-most concept.

Combining the Two Modes of Reasoning:

The two modes of reasoning described earlier in this section are combined for concept recognition in an OM that, in general, contains some nodes representing complex concepts. Reasoning happens in a bottom-up manner. The elementary concepts that can be recognized with observation of media patterns are evaluated with belief propagation in the corresponding Bayesian networks. Some of these concepts will, in general, be the lowest level constituent concepts for some complex concepts. These complex concepts are now evaluated using the fuzzy reasoning method. The complex concepts are leaf nodes in some other Bayesian networks that help in recognizing still higher-level concepts. These nodes are treated as media property nodes in these Bayesian networks and are instantiated if the posterior belief values of the nodes exceed a threshold. The process is recursively followed till the belief value of the root node, representing the primary concept, is computed. Thus, MOWL combines probabilistic and fuzzy reasoning modes for concept recognition.

Algorithms 3 and 4 show the steps of the mixed mode of reasoning for

---

Algorithm 3 Reasoning with the OM to recognize a concept in a multimedia document. Part 1.

Inputs: a) Observation Model Ω_{C_p}
        b) Multimedia Document $\mathcal{D}$
        c) Threshold θ for recognizing concept C_p

Output: Boolean value stating whether concept C_p is recognized in $\mathcal{D}$

1: **procedure** MAIN
2:     $Belief(C_p) \leftarrow$ an *apriori* probability value
3:     Compute $\mathcal{S}_n = \{$ n : n is a complex concept node in Ω_{C_p} $\}$
4:     **for** $i = 1$ to $|\mathcal{S}_n|$ **do**
5:        $Belief(n_i) \leftarrow$ProcessSpatioTemporal(n_i, $\mathcal{D}$)
6:        **if** $Belief(n_i)$ is TRUE **then**
7:           Instantiate node n_i
8:        **end if**
9:     **end for**
10:     Compute $\mathcal{L}_n = \{$ n : n is a leaf-node in Ω_{C_p} $\}$
11:     **for** $i = 1$ to $|\mathcal{L}_n|$ **do**
12:        **if** n_i is a not a Complex concept node **then**
13:           $Belief(n_i) \leftarrow$RunMediaPatternDetector(n_i, $\mathcal{D}$)
14:        **end if**
15:        **if** $Belief(n_i)$ is TRUE **then**
16:           Instantiate node n_i
17:        **end if**
18:     **end for**
19:     Propagate Belief in Ω_{C_p}
20:     $Belief(C_p) \leftarrow$ Posterior belief of C_p
21:     **if** $Belief(C_p) \geq \theta$ **then**
22:        return TRUE
23:     **else**
24:        return FALSE
25:     **end if**
26: **end procedure**   ▷ *Continued in Part 2 ...*

---

concept recognition. The inputs to this reasoning process are the OM Ω_{C_p} of primary concept C_p that is to be recognized, the multimedia document $\mathcal{D}$, and a threshold θ, which is the minimum belief value for presence of C_p in $\mathcal{D}$. For each complex node in the OM, a special procedure is called to detect its presence. This procedure involves decomposing it into its component concepts, running media detectors to check if they exist in the document, and then computing whether the fuzzy spatio-temporal relation between the component concepts is manifested in the document. Then each leaf node in the OM is processed. Media detection tools can be run to detect the media patterns specified in the OM.

If detection of the media pattern node or the complex concept node returns a TRUE value, then the node is instantiated in the Bayesian network. Once all leaf nodes have been processed, belief propagation takes place in the OM, and posterior probability at the root node is computed. If belief at the root node is above the threshold θ, then the algorithm returns a positive for the presence of the concept in the multimedia document. The posterior belief at the root node can also be used as a relevance score of the multimedia document for compliance with a query that could be mapped to the primary concept.

Algorithm 4 Concept recognition algorithm. Part 2.

27: **function** PROCESSSPATIOTEMPROAL(n, $\mathcal{D}$)
28:     $Belief(n) \leftarrow 0$
29:     Expand the ST definition of n to get subject s, object o, and predicate ST
30:     $Belief(s) \leftarrow Belief(o) \leftarrow 0$
31:     **if** s is a Complex concept node **then**
32:         $Belief(s) \leftarrow$ ProcessSpatioTemproal(s, $\mathcal{D}$)
33:     **else if** s is a Media pattern node **then**
34:         $Belief(s) \leftarrow$ RunMediaPatternDetector(s, $\mathcal{D}$)
35:     **end if**
36:     **if** o is a Complex concept node **then**
37:         $Belief(o) \leftarrow$ ProcessSpatioTemproal(o, $\mathcal{D}$)
38:     **else if** o is a Media pattern node **then**
39:         $Belief(o) \leftarrow$ RunMediaPatternDetector(o, $\mathcal{D}$)
40:     **end if**
41:     **if** $Belief(s)$ is TRUE & $Belief(o)$ is TRUE **then**
42:         $Belief(n) \leftarrow ComputeFuzzy(ST, D)$
43:     **end if**
44:     return $Belief(n)$
45: **end function**
46: **function** RUNMEDIAPATTERNDETECTOR(MD, $\mathcal{D}$)
47:     **if** Media pattern is detected by the tool which implements the procedural specification **then**
48:         return TRUE
49:     **else**
50:         return FALSE
51:     **end if**
52: **end function**

5.9 Reasoning Modes with Bayesian Network and MOWL

Reasoning in Bayesian network essentially happens through *belief propagation* across the connected nodes. In general, a Bayesian network supports three distinct modes of reasoning [115] depending on the direction of belief flow, as illustrated with an example in figure 5.19. The four nodes in the network represent some physical phenomena,namely cloud, rain, wet grass, and wet road, respectively. Each of the nodes can assume two values, present or absent. The physical phenomena are represented by four binary variables A, B, C, and D, respectively. Each of the variables can assume two states: TRUE and FALSE corresponding to the presence or absence of the entity they represent. In the network, the directed edges mark the causal relation between the nodes: cloud (A) causes rain (B), and rain (B) causes both wet grass (C) and wet road (D). Conversely, we can say that rain (B) is an evidence for cloud (A), and that both wet grass (C) and wet road (D) are evidences for rain (B). We illustrate the three modes of reasoning on this network as follows:

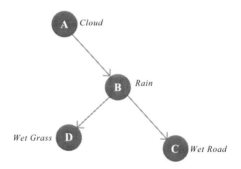

FIGURE 5.19: Illustrative Bayesian network.

1. Causal reasoning: Observation of A will result in belief propagation and change the belief in the state of B. Further, a change in belief in the states of B in turn results in a change in belief values of C and D as well. This type of reasoning is known as causal reasoning, since the belief propagation happens downstream on causal chains. The physical interpretation of such belief propagation is that observation of cloud increases the belief in rain, which in turn increases the belief in wet grass as well as wet road.

2. Abductive (evidential) reasoning: Observation of node C (or D) results in the change in belief value of B, which in turn results in a change in belief value in A. This is called abductive or evidential reasoning as belief propagation happens upstream on a causal chain. The physical interpretation of such belief propagation is that observation of wet grass (or wet road) reinforces our belief in rain (a possible cause of both wet grass and wet road), which in turn reinforces our belief in cloud (a possible cause of rain).

3. Mixed mode reasoning: Mixed mode reasoning combines causal and abductive reasoning patterns in any order. For example, observation of node D changes belief of B (abductive reasoning) and the change in belief value of B in turn results in a change in belief value of C (causal reasoning). The belief propagation happens both upstream and downstream the causal chains. The physical interpretation of such belief propagation is that observation of wet road reinforces the belief in rain, which in turn causes the belief in wet grass to increase.

In section 5.8, we discussed use of the abductive reasoning mode for concept recognition with observation models constructed from an ontology expressed in MOWL. The causal associations in the ontology representation, however, can be used in a more general way. Over and above abductive reasoning, it also provides opportunity for causal and mixed modes of reasoning. Such reasoning modes can be exploited in other contextual inferencing tasks.

5.10   Conclusions

We have described the principles of perceptual modeling and presented MOWL ontology representation language that can be used to realize such perceptual models. The language provides the capability of associating media properties of the ontology concepts with a causal model and with uncertainties. The media patterns can be described in many ways at various levels of abstraction. Complex concepts are constructed by relating elementary concepts with spatio-temporal relations; a key contribution of MOWL is to provide the capability of a formal yet flexible definition for such spatio-temporal relations. The concept recognition in MOWL uses the principle of abductive reasoning, that is, inferencing by best explanation.

The abductive reasoning scheme proposed in the MOWL framework is weaker than the deductive reasoning employed with conventional ontology schemes but is essential for dealing with inherent uncertainties in the observation of multimedia artifacts. The major advantage of the abductive reasoning system is that it can produce robust results with fewer and uncertain observations. Another advantage of this concept recognition scheme is the separation of the knowledge about media properties of concepts from the underlying collection characteristics. The media properties in the ontology should ideally cover varied example instances of concepts in different media forms. In general, multimedia features may not be restricted to visual and audio properties alone, but may include a wider variety of contextual data, such as textual annotations and sensor data of various kinds. Thus, an OM, when constructed from the multimedia ontology, may exhibit a large degree of redundancy in terms of expected media patterns. In general, it is not necessary to observe all associated media patterns to recognize a concept. A method to choose a suitable observation strategy with a subset of observable media nodes is discussed in a later chapter of this book. This flexibility enables integration of multiple distributed collections under a common conceptual framework.

Chapter 6

Modeling the Semantics of Multimedia Content

6.1 Introduction

Ontological reasoning in the multimedia domain addresses the problem of exploiting information embedded in multimedia assets and making the underlying meaning of the multimedia content explicit. However, the process of attaching meaning to multimedia content is not simple, or even well determined. For example, the meaning of an image is not just determined on the basis of image data but also on the situation or context under consideration. Multimedia Web Ontology Language provides a mechanism to attach semantics to the content by specifying possible content-dependent observables of concrete or abstract concepts. For example, using MOWL we can associate a typical huff-and-puff audio track with the categorical concept of a steam engine. There can be other observable features for a steam engine specified using MOWL. The ontological reasoning scheme of MOWL also facilitates specification of possible contexts for the steam engine. Using these specifications, in a multimedia asset such as video, we can search for the possible occurrence of a steam engine, provided we have the appropriate signal analysis algorithm for detection of the huff-and-puff sound and other features. Feature detectors essentially embody techniques for distinguishing specific types of signal instances. Machine learning techniques can be used for building such classifiers and detectors. These classifiers and feature detectors provide the initiation point for semantic modeling of multimedia content in the context of ontological reasoning. The MPEG-7 standard provides a scheme for specifying such descriptors but does not address the problem of generation of descriptors. These descriptors can encode semantic models at different levels of abstraction. For example, *waterfront*, in the LSCOM vocabulary, can be specified as the corresponding image classifier. This image classifier can be used as one of the leaf nodes in MOWL ontology as a detector of *waterfront* based upon image features. MOWL-based ontology can also use contextual features (like spatial relation with a boat) to specify the concept of *waterfront* at an appropriate semantic level. This is the key distinguishing feature of MOWL, which enables semantic model construction in a hierarchical fashion linking higher-level concepts with low-level multimedia data. In this chapter, we review different approaches proposed in the literature for representing and extracting low-level semantic features of multimedia content and how they can be used for constructing deep semantic representations of the concept using MOWL and its reasoning engine.

6.2   Data-Driven Learning for Multimedia Semantics Extraction

Advancements of storage and computational hardware in recent years have made collection and processing of a large volume of image and video data feasible. Consequently, data-driven techniques for extraction of semantics for multimedia data have become more common. These techniques mostly use the principles of statistical pattern recognition. Features extracted from multimedia data provide numeric or symbolic representation of the multimedia content. Ideally features preserve content semantics and are efficiently computable. We can use statistical learning to map features to the semantic labels. A set of labeled examples can be used for generation of a classifier, which, based upon extracted features, can classify multimedia elements into the class of the desired label. Support vector machines (SVMs) [195] are one of the most well-known learning algorithms. An SVM is preferred for its good generalization performance and sound theoretical justifications compared with other algorithms. SVMs are built on the principle of structural risk minimization. They seek to find a decision surface to separate the training data into two classes having maximal margin between them. Apart from SVMs, a large variety of other classifiers have been used in the mutimedia domain. These are Gaussian mixture models (GMM), K-Nearest Neighbor (KNN) [148], Adaboost [117], and random forests [23].

Graphical models provide natural representation for data streaming over time and space. One of the most well-known graphical models is Hidden Markov Models (HMMs) [161]. It is popular because of its simple structure and availability of efficient inference and learning algorithms. If graphical models are combined with discriminant models, it can improve the detection of semantic concepts. Conditional random fields have also been used for learning spatial and temporal concepts [23]. Supervised learning-based approaches are simple and effective, but they do not make full use of the latent structures and inherent data distributions. The idea of latent structures and hidden topics was first explored in text retrieval. There, each document d in collection D consists of words w in vocabulary W. The text data in a collection are summarized into a feature matrix MDW containing the word counts for each document. This is known as the Bag of Words representation, with $D = |D|$ and $W = |W|$. The algorithms then finds K latent topics to best represent

M. Latent semantic indexing (LSI) [50] considers each document as a linear combination of latent topics. There are a few probabilistic extensions to LSI. Probabilistic latent semantic indexing (pLSI) [93] expresses the joint probability of a word w and a document d as

$$p(d, w) = p(d) \sum_z p(w \mid z) p(z \mid d) \qquad (6.1)$$

with z the unobserved topic variable, and $p(z|d)$ taking the role of the topic mixing weights. The latent Dirichlet allocation (LDA) model (developed by Blei et al. [19]) offers even more flexibility by modeling the top-mixing weights as random variables following prior distributions.

One way for concept detection in multimedia documents is to jointly model the associations between annotated concept words and multimedia features. This scenario has wide applications. For example, an image and text data model fits many real-world image collections: professional stock photo catalogs, personal pictures, images on the web with surrounding HTML text, or images on media-rich social sites such as Flickr and Facebook. These approaches typically assume that words and image features are generated by one set of hidden information sources, which characterizes the hidden semantics [86].

Supervised learning for multimedia concept and features extraction suffers from other problems as well. A very common problem is unbalanced training data. Multimedia collections usually contain a small fraction of positive examples. Many learning algorithms like SVM generate inappropriate classifiers. When the class distribution is skewed, SVMs will generate a trivial model by predicting everything to the majority class. One way to overcome this problem is to use oversampling, which replicates positive data. Another approach involves under-sampling which throws away part of negative data. However, both undersampling and oversampling suffer from known drawbacks. Undersampling is likely to eliminate some of the potentially useful examples, and such loss of information may hurt the classification performance. Oversampling, on the other hand, may lead to overfitting and an increase in the learning time.

As an alternative to modifying skewed data distribution, ensemble-based approaches have been proposed in recent studies, of which the basic idea is to combine multiple individual classifiers on balanced data distributions. Modified learning algorithms for SVM have also been suggested in the literature [107]. Generation of a large annotated training corpus for learning is a difficult problem. In order to overcome this problem, a variety of semi-supervised algorithms have been developed to make use of unlabeled data in the training collection. Multiple modalities in multimedia data — in particular, video — can be processed through multiview learning strategies. In multiview learning the feature space is split into multiple subsets, or views. Multiview semi-supervised learning provides a powerful technique for learning with unlabeled video. Co-training [20] is one of the most well-known multiview semi-supervised learning algorithms. Initially, two classifiers are learned from separate feature spaces. Both classifiers are then incrementally updated iteratively using an augmented labeled set, which includes additional unlabeled samples with the highest classification confidence in each view. In another approach, called semi-supervised cross-feature learning (SCFL) [227], unlike co-training, separate classifiers are learned from selected unlabeled data and they are combined with the classifiers learned from noise-free labeled data. Even when views are not sufficient, this approach prevents significant degradation of the classifiers. Another approach is developed using active learning. In active learning, a set of unlabeled data examples is manually labeled positive or negative and classifiers are repetitively learned from newly obtained knowledge. The effectiveness of active learning for semantic concept detection has been illustrated in [33, 139, 226, 189].

6.3   Media Features for Semantic Modeling

In chapter 4, we introduced the MPEG-7 standard, which provides specifications of a set of features for characterizing multimedia content. In the following sections we briefly summarize commonly used features for semantic characterization of image, video, speech, and audio content. More details are available in the survey papers [71, 110, 178, 29].

6.3.1   Image-Based Features

Different content-based features have been used for semantic categorization of still images. Typical features are based upon color, texture, and shape [178]. Color is the most important cue for human beings to distinguish between images. Most systems using color features utilize color spaces such as Hue-Saturation-Value (HSV) [186] and generate a color-histogram representation of the image. Different similarity metrics have been used for comparing histograms. This technique can be further extended by modeling spatial correlation of color pixels in form of Spatial Chromatic Histogram (SCH) [35]. Variations in image illumination in real-world images have been taken care of by using a color-constancy-based algorithm [187]. Texture features represent local patterns in the image. Haralick et al. [82] proposed elementary texture features in the form of a gray-level co-occurrence matrix. Statistics extracted from the co-occurrence matrix have been used as features. Gabor-transform-based features representing texture in spatial and frequency domain have also been used [158]. Markov models and analysis tools have also been found to be useful for handling texture features. One of the earliest systems, the IBM QBIC system [59], used shape features for image retrieval. These features included shape area, eccentricity, and major axis orientation. Geometric features invariant to Euclidean, affine, and projective transformation have been used for shape recognition [124]. The Shape Context feature measures shape similarity through point correspondences in uniform log-polar space [136].

Most popular features used in recent times for capturing local appearance of objects and images are SIFT (and its variants) [124] and HoG [41]. SIFT (Scale Invariant Feature Transform) feature vectors are invariant to image translation, scaling, and rotation. These feature vectors are partially invariant to illumination changes and robust to local geometric distortion. Feature vectors are computed at key locations, and key points are defined as maxima and minima of the difference of Gaussians function in scale space (defined by different sigma values of the Gaussian). A window of size 16×16 is considered around the interest point at the scale detected. This window is divided into a 4×4 grid of cells. Then a histogram of image-gradient directions in each cell is computed to generate a 128 (16 histograms $\times 8$ directions) dimensional feature vector. Computation of Histogram of Oriented Gradeints (HoG) feature vector requires finding gradients at each pixel in a given cell. Each pixel within the cell casts a weighted vote for an orientation-based histogram based on the values found in the gradient computation. As for the vote weight, pixel contribution is typically the gradient magnitude itself. In order to take care of variations in illumination and contrast, the gradient strengths are locally normalized for cells grouped into larger spatially connected blocks. The HoG descriptor is then the vector of the components of the normalized cell histograms from all of the block regions.

These local features are clustered to create a codebook or vocabulary. An image is represented using a histogram of vocabulary counts. Histogram distances are used to compute a similarity score between two images. This method is referred to as the *Bag of Words* (BoW) technique. These histograms are treated as feature vectors for standard classifiers like SVM. This approach has been used for object recognition in [40]. BoW feature vectors are clustered over image collections for discovering visual themes. Hierarchical Bayesian models for documents (pLSA, LDA, etc.) based upon BoW features have been used for object categorization [40, 176, 184] and natural scene categorization [198, 55, 22]. Correlograms of words have been used for incorporating spatial information in BoW features [169]. In order to explicitly model location information and handle spatial relations, objects or scenes have been modeled as a set of parts and relative location between parts. Generative models have been learned for object classes. Issues involved are (i) How to model location; (ii) how to represent appearance; and (iii) how to handle occlusion or clutter. Given an image of an unknown object, correspondence between parts defined in the model and those of the image has to be established as the first step. However, this requirement can lead to exponential complexity. Using distance transform, an efficient approach was proposed in [56]. In [57] SVM-based robust part detectors using HOG features have been used. Use of a distance transform enabled examination of every image location for instantiation of parts in an efficient fashion. A hierarchical compositional system

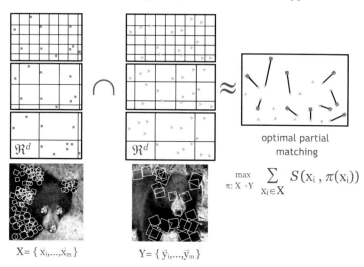

FIGURE 6.1: Kernel feature match.

for rapid object detection has been proposed in [223], in which an object is represented through a hierarchy of parts, and parts are learned at each level. A different approach for a learning cascade of part-based boosted classifiers for object detection has been proposed in [196]. In this work, local features are extracted by Haar wavelet filters. This has become the defacto standard for face detection in images.

Recognizing the category of an object is more complex than recognizing the object itself. But category descriptor is a fundamental requirement for generating the semantic label of an object. Category recognition may require comparing a large number of local feature sets, where each set can vary in cardinality, and the correspondence between the features across sets is unknown. Such a situation occurs, for example, when objects are represented by a set of local descriptors defined at each edge point. In [72], the pyramid match kernel, a new kernel function over unordered feature sets, has been proposed. This allows unordered features to be used effectively and efficiently in kernel-based learning methods. Each feature set is mapped to a multiresolution histogram preserving the individual features' distinctiveness at the finest level. The histogram pyramids are then compared using a weighted histogram intersection computation, which defines an implicit correspondence based on the finest resolution histogram cell where a matched pair first appears. Effectively, feature space partitions serve to "match" the local descriptors within successively wider regions (refer to figure 6.1) [72].

Convolutional Neural Networks (CNN) are variants of multilayer perceptrons. These networks were inspired from Hubel and Wiesel's early work on a cat's visual cortex [96]. Each neuron is sensitive to small subregions of the input space, called a receptive field. These neurons are tiled so as to cover the entire input field. These filters are local in input space and are thus better suited to exploit the strong spatially local correlation present in natural images. The network is constructed in such a way that neurons at each layer compute convolution of the output of neurons of the previous layer. After each convolutional layer, there may be a pooling layer. The pooling layer takes small rectangular blocks from the convolutional layer and subsamples it to produce a single output from that block. There are several ways to do this pooling, such as taking the average or the maximum, or a learned linear combination of the neurons in the block. Finally, after several convolutional and max pooling layers, the high-level reasoning in the neural network is done via fully connected layers. A fully connected layer takes all neurons in the previous layer (be it fully connected, pooling, or convolutional) and connects them to every single neuron it has. These convolutional networks have been extensively used for learning feature extractors from input image data for semantic labeling of image inputs [118].

6.3.2 Audio Features

Audio signals are characterized by a number of perceptual features such as loudness, pitch, and timbre. Loudness is modeled in terms of the signal energy distributed over different frequency bands. Simple operations like autocorrelation can be used for pitch detection of monophonic tonal signals. Relative strength of different harmonics is used for computing the timbre of audio signals. A number of simple features in the time or the STFT (Short Time Fourier Transform) domain can be extracted for characterizing everyday sound. More elaborate and robust features detectors have been proposed in the literature for pitch extraction [132, 45], LPC (linear prediction coefficients) [162], and spectral envelopes such as MFCC (mel-frequency cepstral coefficient) [71]. Due to inherent variability and heterogeneity in the properties of audio signal, it is generally modeled as a non-stationary random process. For example, an audio recording of a street scene can include vehicle sounds — sirens, human speech, footsteps, music, and animal/bird sounds — leading to a complex composed signal. Time-frequency analysis over a hierarchy of scales provides a mechanism to capture semantic features of audio. Experiments in the field of psychoacoustics have shown that the human auditory system, in particular the cochlea in the inner ear acts as a bank of overlapping band-pass filters. The center frequencies of the filters are distributed on a log scale, and the bandwidth of these filters (the critical bands) are narrow for low frequencies and significantly wider at higher frequencies. Interestingly, this allows for higher time resolution for high-frequency components. Based upon these empirical principles, features for characterizing audio signals are the magnitude of frequency components within a critical bank of an auditory filter bank. Specifically, the center frequency and the bandwidth of such an auditory filter bank follow a bark frequency scale where the auditory spectrum is divided into 24 overlapping band-pass filters [95].

However, MFCCs have shown consistently good performance for classification and recognition of audio signals. MFCCs were originally used for speech recognition. In music, the pitch range is much wider than in speech, and for high pitches the MFCC has certain limitations. The use of MFCCs in music signal processing is, therefore, not well justified. Wavelet-like features using a 2D Gabor function that results in a multiscale representation have been shown to perform well in speech/nonspeech discrimination. Audio classification systems can easily deal with a significantly larger number of acoustic categories. Notable examples include [218] and [80]. Here the audio classifiers are trained to recognize categories such as animals, bells, crowds, female, laughter, machines, male voices, percussion instruments, telephone, water sounds, and so forth. The main assumption here is that an audio clip is homogeneous and, therefore, the content belongs to only one of the predefined categories. In complex audio scenes, humans tend to identify separate streams and sound sources. In other scenarios — in particular, in musical compositions — the sound sources tend to blend. Most of the audio categorization schemes use a global approach. An alternative approach organizes audio clips and audio categories in a meaningful hierarchy. In [233], Zhang and Kuo present a system of hierarchical categories such as silence portions and portions with and without music components. Liu et al. have designed a system [125], that groups sources into speech, music, or environmental sounds. In this scheme, a given segment is first labeled as speech or nonspeech. Speech segments are classified according to the speakers, while nonspeech segments are further classified as music, environment sound, or silence. Another example in music is provided in [53] where musical instruments and groups of instruments are represented in a taxonomical hierarchy.

6.3.3   Textual Features

Textual data is another source of semantic information. Typically word counts in a bag-of-words representation, LSI and pLSI features and topic modeling using LDA and its variants are being typically used for representing semantics of textual content. Textual content is also obtained from the transcripts obtained with automatic speech recognition (ASR) or closed captioning (CC). Optical character recognition (OCR) also produces textual content from document images. However, the output of ASR and OCR produces noisy textual data due to errors in the recognition process. Specialized techniques have been developed for modeling content of noisy text [42].

6.3.4   Video Features

In a video, we can extract color, texture, and other local features described in section 6.3.1 for a sequence of frames. Further we can use motion-based features from video. Motion provides information about short-term evolution in video. A 2D motion field can be estimated from image sequences by matching local appearance with global constraints. Motion can be represented using optical flows and motion patterns in specific directions.

6.4 Semantic Classification of Multimedia Content

In the previous section, we discussed different features used for recognition and categorization of media instances. In this section, we examine how these features are used for associating semantic labels with these instances.

Semantic audio classification schemes emulate cognitive aspects of human auditory capabilities. In this process, many systems associate a semantic description with audio inputs and not category label only. Notable examples of such systems are those proposed by Slaney [177] and Barrington et al. [12]. In these approaches, probabilistic inferencing is used for associating terms of the text description with unstructured audio clips. Slaney [177] used mapping of nodes of a hierarchical model in abstract semantic space to the acoustic signal space. These nodes are probabilistically represented as words. Barrington et al. [12] describe a similar approach of modeling features with text labels. These techniques use MFCCs as their acoustic feature. Further, by integrating temporal information, such techniques allow for semantic analysis based on both spectral content and temporal patterns of audio.

In the real visual world, the number of categories a classifier needs to discriminate is on the order of hundreds or thousands. For example, the imageNet challenge[1] is about localization and recognition of objects of 1000 categories. Designing a multiclass classifier that is both accurate and fast at test time is an extremely important problem in both machine learning and computer vision. To achieve a good trade-off between accuracy and speed, an approach was proposed in [63], where a set of binary classifiers is organized in a tree or DAG (directed acyclic graph) structure. At each node, classes are colored into positive and negative groups that are separated by a binary classifier. To achieve high accuracy while being fast, relaxed hierarchy structure [133] is used, where "relaxed" means that each node can ignore a subset of confusing classes. Convolutional neural networks have been used in large-scale object recognition using the ImageNet library [118]. A large, deep convolutional neural network was trained to classify the 1.2 million high-resolution images in the ImageNet into the 1000 different classes. The neural network has 60 million parameters and 650,000 neurons. A very efficient graphic processing unit (GPU) implementation of the convolution operation was used for overcoming the computational load.

In order to solve complex object recognition and localization tasks, a novel approach was proposed in [69]. In this approach, convolutional neural net-

---

[1] http://www.image-net.org/challenges/LSVRC/2014/.

works (CNNs) are used for the generation of bottom-up region hypotheses in order to localize and segment objects. Supervised pretraining followed by domain-specific fine-tuning for final recognition have yielded better recognition accuracy. In [47] a new graphical model has been proposed for large-scale object categorization. This new model allows the encoding of flexible relations between labels. The authors have introduced a new formalism that captures semantic relations between any two labels applied to the same object, namely, mutual exclusion, overlap, and subsumption. This hierarchy is used for developing a probabilistic classification model. The methodology significantly improves object classification by exploiting label relations.

6.5 Use of Ontology for Semantic Classification

Semantic classification requires exploitation of semantic relations between elements in a taxonomical hierarchy. MOWL provides a mechanism for representation and probabilistic reasoning with these semantic relations in a principled fashion. As an example, we examine a way with which we can represent the concept of human action using MOWL representation. We use the following scheme proposed in [55] for detecting human action in images. Usually verbs indicate human actions; the action part is associated with objects related to the action. For example, the verb *riding* associated with *bike* indicates human action *riding bike*; replacing *bike* by *horse* indicates *riding horse*. In MOWL, the node *riding* can have two specialization nodes — *bike* and *horse* — indicating two different actions. We can associate image-based observables to these nodes using the MOWL constructs introduced in chapter 5. Given an image of a human action, many attributes and parts contribute to recognition of the corresponding action.

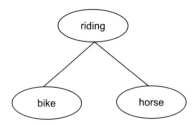

FIGURE 6.2: Concept of "riding" associated with other concepts that help establish context.

Actions are characterized by co-occurrence statistics of objects. For example, the *riding* attribute is likely to occur together with objects such as *horse* and *bike*, but not *laptop*, while the *right arm extended upward* is more likely to co-occur with objects such as *volleyball*. In the scheme proposed in [55], these interactions of action attributes and parts have been modeled as action bases for expressing human actions. A particular action in an image can therefore be represented as a weighted summation of a subset of these bases. In figure 6.2, we show the concept of *riding* (in the limited context) using a MOWL-supported taxonomical structure. The parent node can be represented as the weighted summation of a union of the subsets of children. In fact, an error between reconstruction and test image can be normalized to contribute evidential support. MOWL also provides for specifying observable features for human action in other modalities like text with the same nodes. These features can be used for establishing context with reference to the text associated with an image for a multimodal multimedia document. This discussion shows how different feature-based recognition techniques can be used in an ontological framework supported by MOWL.

6.5.1   Architecture Classification

In this section, we describe an example to show how multimedia ontology can be used for hierarchical category recognition problem. We have chosen the domain of architecture for illustration. The taxonomical structure of architecture domain has been built through scholarly analysis. This concept hierarchy is represented by a MOWL ontology. However, reasoning for classification can only be built upon features detected in a given image. We show how data-driven learning and ontological reasoning of MOWL can be combined for category recognition in a conceptual hierarchy.

For classification of visual scenes into architectural categories, Philbin et al. [156] devised a large scale object retrieval system and demonstrated the system results on an Oxford landmarks dataset. They presented a novel quantization method based on randomized trees to address the problem of complexity in building an image-feature vocabulary of a large dataset. Goel et al. [70] have explored image categorization of European architectures by discovering semantic patterns based upon the repetitive and similar structure elements across monuments. Chu and Tsai [34] exploit a graph mining algorithm to automatically detect and localize repeated structure elements and discover their common feature configurations across monuments for a given architecture. However none of these approaches made use of conceptual hierarchy for reasoning with features detected in the images for categorisation of unknown cases and recognition of learned instances.

6.5.2   Indian Architecture Domain

Among different architectural styles found in India, two distinct schools are Hindu and Islamic architecture. Post 300 BCE, Hindu temple architecture evolved in a distinct way. Particularly, in the period (10th century CE to 13th century CE) monumental temple complexes were built by powerful and wealthy Hindu dynasties. The Hindu temple architecture is typically an open, symmetry-driven structure. It has many variations, involving different geometric shapes. A Hindu temple consists of an inner sanctum, the *garbha griha* or womb chamber, where the primary idol or deity is housed. The *garbha griha* is crowned by a tower-like *Shikhara*, also called the *Vimana*. The architecture includes an ambulatory for *parikrama* (circumambulation), a congregation hall, and sometimes an antechamber and porch. Large temples also have pillared halls called *mandapa*. Mega temple sites have a main temple surrounded by smaller temples and shrines, but these are still arranged by principles of symmetry, grids, and mathematical precision. A possible categorization of Hindu architecture is the following: Nagara, Dravidian, Kalinga, Badami, and Chalukya. A distinctive feature of Nagara architecture is the sharp *Shikhara*. In Dravidian architecture, gate pyramids referred as *Gopurams* are present. Astronomical installations in Jantar Mantar have distinct structural patterns.

Islamic architecture made a sudden but pronounced appearance in north India in the late 12th century. The Qutab Minar complex was started in the late 12th century. It has a mosque and a large open courtyard with pillared ambulatory, massive entrance arches and the disproportionately high *Qutab minar*. Succeeding "sultanate" dynasties continued to build in Delhi up to the Mughal occupation in the mid-16th century. The most prolific were the Tughlaks. Tughlak architecture was characterized by thick sloping walls, a preferred feature in all their secular, military, and religious architecture. The Mughal empire, became the dominant political entity in the late 16th century and stayed dominant till the late 18th century. In these 200 years a large number of imperial and subimperial buildings were commissioned in an evolving Indo-Islamic style of architecture. The Taj Mahal, a tomb built in the mid-17th century at Agra, is considered the pinnacle of Mughal architectural achievement. The common structure found in Mughal tombs is an onion-shaped dome. Minarets or thin spire-like structures are also common. Religious places like mosques also have an onion-shaped dome with a wide entrance gateway. Mughal forts typically have red-colored walls and bastions.

A graphical view of the multimedia ontology of Indian architecture using MOWL representation is shown in figure 6.3. This ontology captures the knowledge about the different styles of Indian architecture in domain concepts and their relations, which are mainly hierarchical. It allows for media properties and media examples (images of specific monuments) to be attached to domain concepts at the lowest-level nodes, which represent the specific sites. It also allows specification of conditional probabilities with concepts at different causal links with other concepts and media properties in the ontology graph.

Top-level concepts in the ontology represent the general architecture styles, like Hindu Architecture or Islamic Architecture. At intermediary levels are more distinct architecture style classes like Tomb, Fort, which are subclasses of Islamic Architecture and Hindu Temple Architecture. Astronomical Instrument, for example, is a subclass of Hindu Architecture. At the lowest level are monument instances like

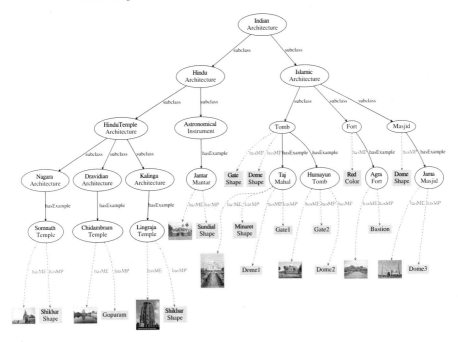

FIGURE 6.3: A multimedia ontology of Indian architecture.

`Tajmahal`, `Somnath Temple`, `Jama Masjid`, and `Jantar Mantar`. As we go down the hierarchy, the architecture elements of a particular style get more distinct, and can be observed as media patterns in images or videos. These observable media patterns can be associated with an architecture style. For example, all tombs have two common shape elements: `Dome Shape` and `Gate Shape`. These are specified with the concept `Tomb`. Monument instances have very specific building elements, which may be distinct from other monuments. For example, the region-of-interest-based discriminators discovered that domes associated with different monuments are distinct. So `Tajmahal` and `Humayun Tomb` have two distinct styles of domes associated with them as media patterns: `Dome1` and `Dome2`. Another dome shape is associated with the monument `Jama Masjid`.

The leafnodes in the ontology graph are media patterns and media examples of monuments. They are associated with respective concepts with *hasMP* and *hasME* relations, which are short for *has Media Pattern* and *has Media Example*. The links that attach them with the concepts are causal and so CPTs can be specified in the ontology accordingly. They are not shown in the ontology figure to avoid clutter. Besides region-of-interest discriminative features that reflect architectural elements, there are other features like color and texture that have been added to some monument nodes in the ontology.

6.5.3 Region Discriminators

Region discrimination techniques are based on the ability of robust local features to discriminate the region of interest from the rest of the image. In this section, we describe how known semantic structural elements (regions of interest) are located in the architectural images using random forests as region discriminators.

(a) `Mughal-Chhatri` characterizing Mughal tombs.

(b) `Shikhar` characterizing `Nagara` style of `Hindu` temples.

FIGURE 6.4: Common architectural styles characterizing classes of monuments.

6.5.3.1 Semantic Features

The human brain has the recognition capability to easily identify architecture styles based on presence and combinations of structural elements in monument images. These architecture styles can be encoded as domain concept nodes at top levels in an architecture ontology. It is difficult to train machine-learning systems to recognize these high-level semantic concepts efficiently. But these systems can be trained to recognize low-level media features like some local discriminative regions within an image. For example, monuments of Islamic architecture generally have *minarets* present, while Chinese architecture generally shows multi-inclined roofs in their monuments. These discriminative regions can be of any size, shape, color, and texture. We use random forests to locate the discriminative regions. Figure 6.4 show some Indian monuments with structural elements that are common to the architectural style to which they belong. These can be used as local discriminative features in region discrimination.

6.5.3.2   Learning Discriminative Region Using Random Forest

The random forest technique [231] is used to identify the image patches that are useful for categorizing. We extend their work on random forest based on discriminative decision trees. They randomly sample discriminative patches for recognition using random forest. In our framework, we sample the regions of interest based on explicit knowledge of the discriminative structure elements. This knowledge is encoded in an ontology of the domain. Discriminative classifiers based on explicit knowledge are applied at the nodes of decision trees to locate the region of interest that gives the best split of data. At the leaf nodes, the posterior probability is obtained as a confidence score of the discriminative region obtained from random forests as shown in algrorithm 5. Given examples of different architectural styles, the learning algorithm identifies distinctive region features that can be used as media descriptors for leafnodes of the MOWL ontology.

Algorithm 5 Random forests with prior knowledge from architecture ontology

---

1: **procedure** RANDOMFORESTWITHKNOW(X_{train},Y_{train})
2:     **for all** Decision tree t **do**
3:         Obtain a random set of training examples X_{train};
4:         Train SVM with ROI from prior knowledge as attributes
5:         SplitNode (X_{train});
6:         **if** needs to split **then**
7:             (i) Randomly sample the candidate ROI in training examples
8:             (ii) Categorize training examples using learned SVM into class 'c' and not class 'c'
9:             (iii) Select ROIs which give best split
10:             (iv) SplitNode (X_1) and SplitNode (X_2)
11:         **else**
12:             Return posterior probability $P_c(t)$ of class 'c' in the current leaf node
13:         **end if**
14:     **end for**
15: **end procedure**

6.5.4   Architecture Categorization Using a Multimedia Ontology

In previous chapters, we saw how the Multimedia Web Ontology Language provides a mechanism to attach semantics to the content by specifying possible content-dependent observables of concrete or abstract concepts. For example, we can associate several observable features like visual shape and typical texture of building material (marble) with the categorical concept of `Tajmahal`. The reasoning scheme of MOWL also facilitates specification of possible contexts for the Taj Mahal, such as region discriminating features for the "dome" shape, texture, or color classifiers. Using these specifications, we can search for possible occurrence of Tajmahal in images, provided that we have appropriate pattern recognition algorithms for detection of the dome shape and other specified features.

As an example, we examine the way we can represent the concept of a monument using an architecture domain ontology. We find that a kind of Indian medieval monument is a "Tomb," which reflects a certain architectural style. Typically two kinds of architectural elements are found in all tombs: "Dome" (onion-shaped domes) and "Gate" (a distinctly shaped entrance to a tomb). Different tombs may have distinct dome shapes and are made with different materials — white marble or red stone. Some of them may have other distinct structural elements like "Minarets". The concept node `Tomb` has sub-concepts `HumayunTomb` and `Tajmahal` indicating two different monument instances. We can associate image-based observables to these nodes using MOWL. Given the image of a tomb, many attributes and parts contribute to the recognition of the corresponding monument.

Monuments are characterized by co-occurrence of structural elements as well as of some color and texture features. For example, "Dome" and "Minarets" are likely to occur together with "Gate" but not with, say, "Gopuram," which is a structural element of a Hindu temple. Similarly structural elements of Hindu temples like "Chhatri" and "Shikhar" are most likely to co-occur. Using this knowledge from the ontology, the feature detectors and classifiers built to recognize concepts in the architecture domain, can be tuned to recognize combinations and co-occurrence of structural elements in image documents. These combinations are encoded through implicit conditional probabilities associated with the links in a MOWL representation. For a given media document like an image, evidence is gathered to contribute combined posterior probability for classification and detection.

6.5.5   Experiments: Indian Architecture Categorization

We evaluated performance of our proposed hierarchical framework on Hindu and Islamic architecture styles. Training of the random forest classifier with region discriminators was done with 1000 images of 12 monuments

FIGURE 6.5: Islamic monuments with region discriminators.

of Indian architecture from Flickr.[2] Images of eight monuments were used in training the framework and tested on the images of remaining four monuments that were not present in training but belonged to some category present in training.

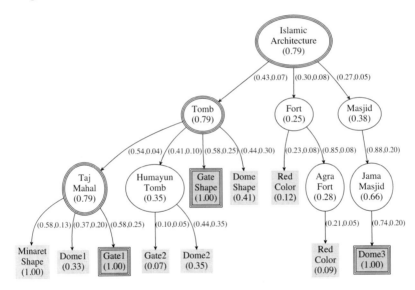

FIGURE 6.6: Islamic architecture observation model showing categorization of an image as `TajMahal`, `Tomb`, and `IslamicArchitecture`.

Using the region-discriminator features as media nodes in the observation model (OM) of concept `IslamicArchitecture` shown in figure 6.6, we show the result of architecture categorization through belief propagation in the OM, which is a Bayesian network. The OM shows the CPTs as a pair of probabilities : $P(C_1 \mid C_2)$ and $P(C_1 \mid notC_2)$ where C_1 is the child of concept node C_2 in the ontology. For an unlabeled image of Taj Mahal, the region discriminators

---

[2]https://www.flickr.com/.

yield observation of Gate shape and Minaret shape with a probability higher than a threshold. The corresponding media nodes are instantiated in the OM, and belief is propagated, which yields high belief for the presence of the nodes Tajmahal, Tomb, and Islamic Architecture. This happens even though all media nodes for the concept Tajmahal were not recognized.

For another test image, which is of ItmadullahTomb, a monument not used in training, the region discriminators yield observation of Gate shape and Dome shape with a probability higher than a threshold, but the Dome shapes that are specific to the monument instances Tajmahal and HumayunTomb in the OM are not recognized or instantiated. On belief propagation, concepts Tomb and IslamicArchitecture get high belief for this monument, and thus it is correctly categorized as a Tomb of the IslamicArchitecture, due to the co-occurrence of structural elements that identify a tomb.

We compared the classification of an SVM classifier trained with region discriminators against the recognition done by reasoning with OMs generated from the MOWL ontology. The posterior probability in the ontology-based inferencing comes from the random forest classifiers that have been trained with the knowledge of structural elements, their combinations, and co-occurrence.

Entity	SVM		MOWL	
	Precision	Recall	Precision	Recall
Low-Level: Structural Elements				
Dome	0.94	0.87	0.92	0.88
Gate	0.90	0.83	0.91	0.81
Minaret	0.98	0.88	0.97	0.89
Intermediate-Level Semantic Concepts: Monuments				
Tajmahal	0.80	0.78	0.90	0.88
Humayun's Tomb	0.73	0.85	0.86	0.89
Jama Masjid	0.81	0.83	0.85	1.00
High-Level Abstract Concepts: Architectural Styles				
Tomb	0.70	0.60	0.84	0.89
Fort	0.47	0.47	0.84	0.84
Masjid	0.44	0.47	0.80	0.87
IslamicArchitecture	0.29	0.36	0.72	0.88

FIGURE 6.7: Comparative results of architecture categorization using SVM, and MOWL ontology.

Figure 6.7 shows a table with the results for categorization of some of the concepts from the Indian architecture with a focus on Islamic architecture style. We find that while results for SVM and MOWL framework are similar for low-level structural elements, they improve for the categorization based on the latter for intermediate levels, which are the monuments. Ontology-based reasoning is able to identify the architectural style for images of *unknown* monuments as well, due to observation of low-level structural elements and their combinations, whereas SVM classifier fails to give good results in such cases. Thus high-level semantic concepts that represent the architectural styles like `Hindu` and `IslamicArchitecture`, are also well classified by the architecture ontology-based reasoning framework due to the knowledge encoded in the ontology. Thus these experiments illustrate the effectiveness of this framework, which combines ontological knowledge with machine learning of media features to successfully categorize images of buildings in Indian architectural styles.

6.6 Conclusions

In this chapter, we have examined different data-driven feature extraction and machine-learning-based semantic analysis techniques. We have found that data driven techniques are useful for semantic characterization of low-level features and feature agglomerates. But concept association with multimedia content is a more complex problem. An observation model of a concept provides the knowledge-based expectation of content in a multimedia entity. Combination of data-driven concept hypotheses verified through observation model-based reasoning provides a more powerful approach. We concluded the chapter by presenting an example of such an approach for architecture-based classification of images of historic sites.

Chapter 7

Learning Multimedia Ontology

7.1 Introduction

Progress in artificial intelligence has always led people to believe in *intelligent* machines. But software agents that are expected to replicate human intelligence are typically knowledge-based systems. These systems are built with domain knowledge acquired from domain experts and other information sources. As these systems grow in size and become more complicated, acquiring the knowledge to build them becomes a challenge. In this context, automating the process of knowledge generation becomes a desired objective. The World Wide Web consortium, which has been working to formulate standards for an *intelligent web* or Semantic Web, has presented ontologies as an ideal technology for representing knowledge. An ontology encodes a shared vocabulary that can be used for modeling a domain. This vocabulary contains the terms of the domain, mainly the types of *concepts* that represent the domain, their properties, and the relations between them. In fact, it has been established that ontologies are an excellent medium for capturing domain knowledge. An ontology can also be used to reason about the concepts and properties of that domain to generate new knowledge about the domain, if required. Thus a domain ontology (or domain-specific ontology) successfully models a specific domain, or part of the world.

Modeling a domain and building ontologies manually is a challenging and time-consuming task. Manually constructed ontologies are error prone, costly, unadaptable, and mostly reflect the bias of the developer. Thus there is sufficient incentive to automate the process. Ontology learning is defined[1] as the automatic or semiautomatic creation of ontologies, including extracting the corresponding domain's terms and the relationships between those concepts from a corpus of natural-language text, and encoding them with an ontology language for easy retrieval. Terms equivalent to *ontology learning* are *ontology acquisition*, *ontology extraction*, and *ontology generation*. Building an ontology requires employment of techniques like data mining and machine learning from complementary disciplines like knowledge discovery and artificial intelligence. Ontology learning can be applied to closed domains like medicine and cultural

---

[1] http://en.wikipedia.org/wiki/Ontology_learning

121

heritage, where data is located in medical databases and digital heritage collections. It can be extended to open domains such as the web, where learning is done from text and multimedia data embedded in the web documents.

Ontology construction is necessarily an iterative process. An ontology representing concepts and relationships of the domain can be constructed manually with a domain expert providing the inputs. But the building of such static and hand-crafted ontologies requires knowledge acquisition from domain experts, which in itself is a complicated task. Extensive interactions are needed between experts and the ontology engineers, and there is still the possibility of missing some concepts and relations that may exist in the real-world, or of coding some extra knowledge that might be obsolete. It is more effective to fine-tune the knowledge obtained from the expert by applying learning from real-world examples belonging to the domain. An ontology refined in this manner is a better structured, logically valid model of the domain that it represents. The goal of ontology learning, thus, is to combine knowledge discovery with machine learning to (semi-)automatically generate relevant concepts and relations from a given corpus and expert inputs.

In the previous chapter, we discussed techniques for semantic analysis of multimedia data based on media feature extraction and machine learning. There we illustrated with an example how ontology-based domain knowledge combined with data-driven feature learning, leads to a more powerful approach to multimedia access and retrieval. Now we discuss ontology learning techniques with a focus on multimedia data space. Most of the current approaches in this area focus on learning ontologies from text – in content or in metadata attached to multimedia. Some of these techniques are detailed in section 7.2. Although these techniques may work for ontologies in linguistic space and to some extent in multimedia space, a different approach is required for effectively learning ontologies from multimedia data. We present a multimedia content management framework in section 7.4 that incorporates one such approach for multimedia ontology learning. In this framework, knowledge of the domain experts is used not just to build an ontology but also to provide semantic metadata for media documents, which helps in fine-tuning the parameters of a MOWL-based ontology. Further sections of the chapter elaborate on the process of learning and establish how a multimedia ontology of the domain learned in this way leads to much improved semantic access to media content.

7.2   State of the Art in Ontology Learning

The field of ontology learning is extensive and much research has happened over the years. The initial trend was to use human inputs to build *seed ontologies* that were then refined using automated ontology learning tech-

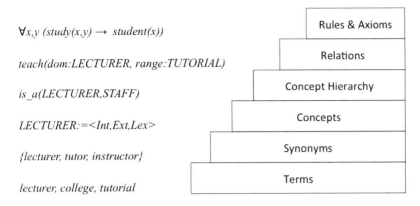

∀x,y (study(x,y) → student(x))

teach(dom:LECTURER, range:TUTORIAL)

is_a(LECTURER,STAFF)

LECTURER:=<Int,Ext,Lex>

{lecturer, tutor, instructor}

lecturer, college, tutorial

Rules & Axioms

Relations

Concept Hierarchy

Concepts

Synonyms

Terms

FIGURE 7.1: Ontology learning layer cake.

niques. Over the years, the trend has changed towards using semiautomated and fully automated techniques to learn ontologies directly from data. Buitelaar and Magnini [25] proposed that the complex task of learning of an ontology from data can be broken down into subtasks shown in figure 7.1 as an *ontology learning layer cake*. The subtasks include learning different components of an ontology which are *terms, synonyms, concepts, concept hierarchies, non-hierarchical relations*, and *rules*. Natural-language processing (NLP) techniques are required to derive *terms*, which are the words or phrases that define most relevant events and objects of a domain. Thesauri and dictionaries are needed to expand these terms with synonyms. Learning domain concepts that are more abstract than terms requires more advanced data-mining techniques. Thus as we go up in the layer cake, the tasks become more complex, requiring techniques that employ machine learning and artificial intelligence.

The complexity of ontology learning process depends upon the kinds of data available from which the learning has to take place. Ontologies can be learned from *unstructured data* which could be natural text from books, articles, and so on; *semistructured data* like HTML and XML documents; and *structured* data like databases and dictionaries. Different techniques exist for learning an ontology from different kinds of data. Statistical analysis and natural-language processing (NLP) approaches are mainly used separately or in tandem for building ontologies from free (unstructured) text. Data-mining techniques like clustering are best used with semi-structured data. Web content mining approaches are used for mining ontological terms and concept hierarchies from HTML document sources, Document Object Model (DOM) structures, and so on, An overview of the state of the art in ontology learning, evaluation methods and challenges faced by the researchers in the field is provided in [236, 88, 120]. Another study [126] has focused on learning ontologies for the Semantic Web. According to all these surveys, text is the most used medium for learning ontologies.

Another way to categorize the techniques for learning an ontology from text is by the kind of ontology being learned. In [220], the authors have broadly classified ontologies as

1. Controlled vocabularies like glossaries, data dictionaries, and web directories, which have defined and regulated terms.

2. Taxonomies like XML DTDs, which are controlled vocabularies with hierarchical structures.

3. Thesauri like the WordNet [135], which are extensions of taxonomies with additional non-hierarchical relations.

4. Lightweight ontologies like data models, XML Schemas, and formal technologies, which do not make use of axioms.

5. Formal, heavyweight ontologies like Frames and Description Logic-based ontologies which make use of axioms and rules.

Most of the automated ontology learning systems that claim success are in fact creating *lightweight* ontologies from text [220]. The ontology learning techniques can broadly be classified as *statistics based, linguistics based, logic based,* and *hybrid.* Statistics-based techniques include clustering [219], latent semantic analysis [194], and co-occurrence analysis [24]. Some of the linguistics-based or NLP techniques are *part-of-speech tagging, parsing sentences,* and *analysis of syntactic structure or dependencies.* These are common techniques used in most of the systems that learn ontologies from text. Logic-based approaches are needed mainly for learning axioms for heavyweight ontologies. They involve reasoning and machine-learning techniques like logical inference [174]. An example of logic-based ontology learning can be found in the OntoLearn system [140], which extracts relevant domain terms from a corpus of text, relates them to appropriate concepts in a general-purpose ontology, and detects taxonomic and other semantic relations among the concepts. Progressively the trend is towards using hybrid approaches that comprise many of the above-mentioned techniques. A relatively new approach is *Crowd-sourcing* ontologies which combines automated learning with human inputs. An example is the use of the Amazon mechanical turk[2] for constructing taxonomies. There is increasing interest in learning ontologies from social data available on the web like in blogs, wikis, folksonomies, and sites like citeulike.com and imdb.com [188, 116]. Some of the key issues that will drive further research in ontology learning from text are issues of noise, provenance and validity of data, testing the correctness of the learned ontologies, constructing more formal ontologies from existing lightweight ontologies, and building social and cross-language ontologies.

---

[2]en.wikipedia.org/wiki/Amazon_Mechanical_Turk.

7.3   Learning an Ontology from Multimedia Data

The knowledge is often encoded in multimedia format. For example, video recordings of performances by renowned artistes are generally used as examples for training purposes. The images of murals and sculptures, and videos of rituals, festivals, traditional sports, and so forth, constitute knowledge-base in cultural heritage domain. In such domains, it may be extremely difficult to express the knowledge in the textual form. In such domains, the knowledge to build an ontology needs to be extracted from multimedia examples. The challenges of learning an ontology from multimedia data are far different and much more complex than those for constructing an ontology from text data. The data that is available with multimedia collections is of two kinds: (a) *Content-based features* extracted from the multimedia data, and (b) *Textual metadata* or annotations that give additional knowledge about the content. While text can be processed using NLP techniques, multimedia content needs special feature-extraction tools to process and analyze it. Additionally multimedia content has contextual information, as well as spatial and temporal relations between its components and segments. Multimedia content-analysis techniques may yield media-specific information about the content of the documents but have limited success in extracting its semantics. For instance, image processing techniques can yield information about the color, texture, and shape of objects in image documents but not about the event, theme or product that they depict. Figure 7.2(b) shows a scene near the sea, but image processing for color features can only yield that the dominant colors in the image are blue and green. Text data mining approaches like latent semantic indexing (LSI) or topic modeling, when applied to these media features can at best generate media-specific clusters or topics but provide little information about the domain. This is the classic *semantic gap* issue between low-level media-features and high-level semantic concepts.

Earlier in chapter 4, we discussed different kinds of metadata associated with images — content-independent metadata like names of authors and dates, content dependent metadata related to low-level media features, and descriptive or semantic metadata. Another option to acquire domain knowledge from media data is through this metadata, or text annotations accompanying media documents, that can be processed and analyzed to help construct ontologies. But there are issues with this approach. Media annotations are often inadequate, inaccurate, and subjective. Almost 70 percent of images and videos on social media are not annotated. Annotations often describe context, not content. This can be seen from figure 7.2, where the annotations describe the events but say nothing about the content. Media data can have multiple interpretations and users perspectives can be different from those of the authors of annotations. In spite of these challenges, text annotations and metadata are a useful source for acquiring semantics of multimedia data. Techniques

(a) Maya's 6th Birthday   (b) June Goa Holiday

FIGURE 7.2: Image annotations: depict context not content.

for learning ontologies from text can be applied to this data to construct ontologies.

We have found that limited research exists in the area of ontology learning from multimedia data. One of the initial research projects that tried to bridge the semantic gap by a collaborative effort to define a standard formal vocabulary for the annotation and retrieval of video was the Large-Scale Concept Ontology for Multimedia (LSCOM) project [138], mentioned earlier in section 4.5. The effort led to formation of a LSCOM taxonomy of 2,638 concepts, built semi-automatically, and later to the full LSCOM ontology, in the form of a 2006 ResearchCyc release that contains the LSCOM mappings into the Cyc ontology [215]. The LSCOM project focused more on developing a concept taxonomy, rather than a full-fledged heavyweight ontology with inference rules. Its main applications are in the areas of concept-detection and video annotation [180, 179]. In another project [18], the authors used logic-based techniques to automatically learn rules describing high-level concepts. They exploited the domain knowledge embedded in an ontology to learn a set of rules for semantic video annotation, using an adaptation of the First Order Inductive Learner (FOIL) technique to the Semantic Web Rule Language (SWRL) standard.

Methodologies that combine multimedia content analysis and processing of the textual metadata and utilize them to complement each other, have better success in building ontologies from multimedia data. One of such approaches is used in [105], where multimedia ontologies are constructed semiautomatically from a video collection by combining video-analysis tools with text processing of the video annotations. In section 7.4, we describe one such framework where both kinds of data are used to refine a basic (seed) multimedia ontology of the domain. The multimedia ontology, which is the backbone of this framework, is constructed with the help of knowledge obtained from domain experts and then further fine-tuned by learning from real-world data that are parts of the digital resources of the domain.

7.4 Ontology-Based Management of Multimedia Resources

Multimedia resources pertaining to a domain can include digital replicas of domain artifacts, events, and so on. Some examples of the artifacts can be videos and still photographs of sports events, scanned images of paintings, video recordings of dance performances, and so forth. The contextual knowledge about these resources contributed by domain experts also form an important ingredient of the multimedia knowledge resources. The framework for multimedia resource management, detailed here, is motivated by the challenge of relating the digital media objects with context-based knowledge, in order to make the data more usable. With this requirement, we discuss this ontology-based framework for multimedia content management with flexible structure and dynamic updation. As shown in figure 7.3, there are four main stages in the framework:

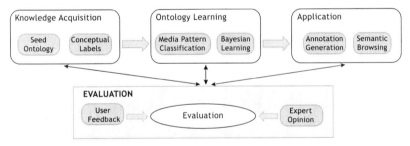

FIGURE 7.3: Framework for ontology-based management of multimedia content.

- **Knowledge Acquisition**: This stage deals with acquiring the highly specialized knowledge of a domain and encoding it in a domain-specific ontology. It also involves collecting the multimedia data of the domain and building a digital collection. To begin with, a basic seed ontology for the domain is hand-crafted by a group of domain experts. The ontology includes the domain concepts, their properties and their relations. The domain experts also provide conceptual labels to a training set of multimedia data. They annotate the multimedia files and their segments, based on their observations, in such a way that the labels correspond to domain concepts in the ontology.

– **Ontology Learning**: At this stage, the basic ontology, enriched with multimedia data, is further refined and refined by applying machine learning from the training set of labeled data. Multimedia Web Ontology Language (MOWL) is used to represent the ontological concepts and the uncertainties inherent in their media-specific relations. The multimedia ontology, thus created, encodes the experts' perspectives and needs adjustments to attune it to the real-world data. Conceptual annotations help build the case data used for applying a machine-learning techniques to refine the ontology.

An important part of the ontology learning stage is the development of media pattern classifiers or detectors that can detect media patterns corresponding to the lowest-level media nodes in the ontology based on the presence of content-based media features. MOWL supports probabilistic reasoning with Bayesian networks in contrast to crisp Description Logic-based reasoning with traditional ontology languages. The joint probability distributions of the concept and the media nodes are computed and a machine-learning technique can be applied to create the probabilistic associations. The technique is applied periodically as new labeled multimedia data instances are added to the collection and the ontology is updated. This semiautomated maintenance of ontology alleviates significant efforts on the part of knowledge engineers.

– **Application**: The multimedia ontology is used for annotation generation for new instances of digital data. A set of media feature classifiers is used to detect the media patterns corresponding to the media nodes in the ontology. The MOWL ontology can then be used to recognize the abstract domain concepts using a probabilistic reasoning framework. The concepts so recognized are used to annotate the multimedia documents. The goal behind building such a framework is to give a novel multimedia experience with ontology-guided navigation to the user seeking to access resources belonging to a digital collection. The conceptual annotations are used to create semantic hyperlinks in the digital collection that along with the multimedia ontology, provide an effective *semantic browsing* interface to the user.

– **Evaluation**: As the multimedia ontology is created and maintained along with the building of the digital collection, each process in this framework is constantly evaluated for integrity and scalability. Users and domain experts are part of the process of updating the knowledge base as new learning takes place and changes happen in the real world.

As this framework is based on a multimedia ontology that is encoded in MOWL, uncertainty specification and probabilistic reasoning have to be supported. The observation models which are a perceptual models of the domain concepts, are Bayesian networks. Learning of a MOWL ontology can be broken down into learning the observation models and then aggregating this learning into the ontology. With this fact in mind, the learning of this MOWL ontology is proposed in terms of Bayesian network learning.

7.4.1 Bayesian Network Learning

Bayesian network learning is a common statistical machine-learning approach. Its use in ontology learning is limited by the lack of support in standard ontology languages like OWL for probabilistic reasoning. Ding and Peng [48] have proposed a probabilistic extension to OWL by using Bayesian networks, but this is limited to textual data. Here we mention some of the research happening in the field of Bayesian network learning. Starting from his tutorial on learning Bayesian networks in 1995 [90], Heckerman has published several works in this field. His research focuses on structural as well as parameter learning in Bayesian networks. Other algorithms and methods of structure learning in probabilistic networks include so-called naive Bayesian network learning, which states that classification is an optimal method of supervised learning in a Bayesian network if the values of the attributes of an example are independent given the class of the example. Webb and Zheng [234] have considered an extension of naive Bayes, where a subset of the attribute values is considered, assuming independence among the remaining attributes. Niculescu et al. [145] have used parameter constraints to learn the Bayesian network.

Bayesian learning has been used in several applications of information retrieval (IR). Neuman et al. have described a model of IR based on Bayesian networks in [144]. In [2], we see the usage of Bayesian learning for neural networks in predicting both the location and next service for a mobile user movement. Town et al. [190] described how an ontology consisting of a ground truth schema and a set of annotated training sequences can be used to train the structure and parameters of Bayesian networks for event recognition. They have applied these techniques to a visual surveillance problem, and use visual content descriptors to infer high-level event and scenario properties. These applications work in generic, open domains where domain knowledge is not specialized, and there is no learning from metadata attached to the videos.

7.4.2 Learning OM: A Bayesian Network

Here we discuss a technique for learning a multimedia ontology expressed in MOWL in terms of learning the observation models that are probabilistic Bayesian networks. The technique attempts to encode the highly specialized knowledge that experts of a scholarly domain have into an ontological representation of the domain, and refine this knowledge by learning from observables in the multimedia examples of the domain. Combination of domain knowledge with example-driven supervised learning for generation of a domain ontology for multimedia retrieval is a unique contribution of this technique. Two inputs are required in the multimedia ontology-learning process in this technique which uses the MOWL encoding for representing the ontology. These are: (a) a basic ontology of the domain which is constructed with the help of knowledge provided by the domain experts, and (b) conceptual annotations by domain experts based on observable parameters in the media files.

A standard Bayesian network (BN) learning algorithm is applied and extended to learn uncertainty between concepts and their media properties. The basic structure of the BN for a concept, which is the starting point of the learning, comes from its OM drawn from the basic domain ontology. The BN is learned using data from the training set of annotated media instances, which provides the case data for learning. Once the OMs are learned, the learning is then applied to update the structure and uncertainties encoded in the ontology. The efficacy of learning can be tested by building applications of annotating, searching and browsing based on the learned ontology and testing for expected improvement in results.

A Bayesian network is characterized by its topology and the conditional probability tables (CPTs) associated with its nodes. The goal of learning a BN is to determine both (a) the structure of the network (*structure learning*), and (b) the set of CPTs (*parameter learning*). An OM modeled as a Bayesian network is in effect, a specification for the concept in terms of searchable media patterns. The joint probability distribution tables that signify causal strength (significance) of the different media properties towards recognizing the concept are computed from the probabilistic associations specified in the ontology. The basic ontology provides a basic structure of the BN with CPTs reflecting the domain experts' knowledge of the domain. The aim is to use the learning algorithm – to refine this structure, which includes (a) discovering new links or relationships between concepts, and (b) removing some obsolete links, i.e. getting rid of some properties or relationships that do not exist in data; – and to learn the parameters of the Bayesian network. The algorithm must take into account the media observables or features that are associated with the concept nodes.

7.4.3   Full Bayesian Network Learning

For this learning scheme, a standard Bayesian learning technique called the Full Bayesian Network learning (FBN) [183] is selected. The algorithm has been extended to learn structure and parameters of the Bayesian networks that correspond to the OMs for concepts in the multimedia ontology. The data for learning come from the training set of videos using the media-based features of the examples that help assign values to the variables in the network.

BN structure learning often has high computational complexity, since the number of possible structures is extremely huge. FBN overcomes the bottleneck of structure learning by *not using* the structure to represent variable independence. Instead, all variables are assumed dependent and a full BN is used as the structure of the target BN. FBN learning uses decision trees

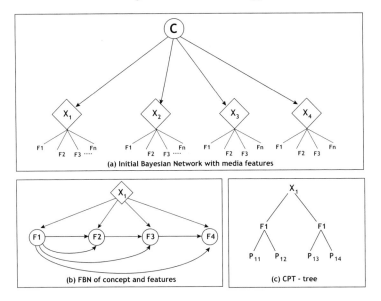

FIGURE 7.4: Full Bayesian network with observable media features.

as the representation of CPTs [61] to obtain a more compact representation. The decision trees in CPTs are called CPT-trees. In learning an FBN classifier, learning the CPT-trees captures essentially both variable independence and context-specific independence (CSI).

For each OM extracted from the base ontology Γ_B, a set of subnets, each of which is a naive Bayesian network, is obtained. The CPTs are copied into each subnet from the OM. FBN learning from case data takes place in each subnet, updating the structure and the parameters of the BN. The learned subnets then update the OM and the learned OMs are used to update the ontology. Figure 7.4(a) shows a subnet that is a naive Bayesian network, constructed from a snippet of the MOWL ontology, showing a concept node C, related to some other concepts X_i by MOWL relations. X_i are further connected to some media nodes shown as leaf nodes F_is in the snippet. These denote the media-observable features associated with the concept. Section 7.4.3.1 explains how FBN structure learning is applied to learn the structure of each subnet; followed by section 7.4.3.2 which explains simultaneous learning of CPTs; and section 7.4.3.3 which extends the FBN learning technique to learn associations of concepts with observable features.

7.4.3.1 FBN Structure Learning

Given a training data set S, it is partitioned into $|C|$ subsets, each S_c of which corresponds to the concept value c and then an FBN is constructed for S_c. Learning the structure of a full BN means learning an order of variables and then adding arcs from a variable to all the variables ranked after it. The order of the variables is learned based on total influence of each variable on other variables. The influence (dependency) between two variables can be measured by mutual information defined as follows:

$$M(X,Y) = \Sigma_{xy} P(X,Y) log P(X,Y) \tag{7.1}$$

where x and y are the values of variables X and Y respectively. Since the mutual information in each subset S_c of the training set is computed, $M(X,Y)$ is actually the conditional mutual information $M(X,Y,|c)$. This ensures a high probability of learning true dependencies between variables. In practice, it is possible that the dependency between two variables, measured by Equation 7.1, is caused by noise. Thus, a threshold value is required to judge if the dependency between two variables is reliable. One typical approach to defining the threshold is based on the Minimum Description Length (MDL) principle. Friedman and Yakhini [62] show that the average cross-entropy error is asymptotically proportional to $log_2 N/2N$ where N is the size of the training data. Their strategy is adopted to define the threshold to filter out unreliable dependencies as follows:

$$\varphi(X,Y) = log_2 N/2N * T_{ij} \tag{7.2}$$

where $T_{ij} = |X_i| \times |X_j|$, $|X_i|$ is the number of possible values of X_i, and $|X_j|$ is the number of possible values of X_j. In structure learning algorithm the dependency between X_i and X_j is taken into account only if $M(X_i; X_j) > \varphi(X_i, X_j)$. The total influence of a variable X_i on all other variables denoted by $W(X_i)$ defined as follows:

$$W(X_i) = \sum_{\substack{j(j \neq i)}}^{M(X_i;X_j)>Q(X_i,X_j)} M(X_i; X_j) \tag{7.3}$$

Once CPTs are learned, as detailed in the next section, the parameters determine whether all the links are retained, or some are deleted.

7.4.3.2   Learning CPT-Trees

After the structure of an FBN is determined, a CPT-tree should be learned for each variable X_i. As per FBN learning, we have used the fast decision-tree learning algorithm for learning CPTs. Before the tree-growing process, all the variables $X_j \in \pi(X_i)$ (parent set of X_j) are sorted in terms of mutual information $M(X_i, X_j)$ on the whole training data, which determines a fixed order of variables. In the tree-growing process, the variable X_j with the highest mutual information is removed from $\pi(X_i)$, and the local mutual information $M^S(X_i, X_j)$ on the current training data S, is computed. If it is greater than the local threshold $\varphi^S(X_i, X_j)$, X_j is used as the root, and the current training data S is partitioned according to the values of X_j and this process is repeated for each branch (subset). The key point here is that, for each variable, the local mutual information and local threshold is computed only once. Whether failed or not, it is removed from X_i and is never considered again. The fast CPT-tree learning algorithm can also be found in [183].

7.4.3.3 Learning Associations of Observables with Concepts

This multimedia ontology learning scheme extends the FBN learning algorithm to learn associations of concepts with observables features. Figure 7.4(b) shows a concept node X_1 with associated media properties $F1$ to $F4$ as its children. An FBN is constructed for each value x_i of X_1 denoting an ordering among media features. CPT-trees denoting uncertainties between a concept and its media properties are learned using the same algorithm as for learning uncertainties between concepts. The basis of the FBN algorithm is the mutual information that denotes the influence (dependency) between two attributes, that is, two media features here. This is computed by Equation 7.4,

$$M(F_i, F_j) = \Sigma_{f_i, f_j} P(f_i, f_j) log P(f_i, f_j) \qquad (7.4)$$

where f_i and f_j are values for F_i and F_j respectively. $M(F_i, F_j)$ is the conditional mutual information $M(F_i, F_j | x_i)$, that is, dependency between the two features, given a value x_i of the concept X_1. To compute $P(f_i, f_j)$, the extracted features need to be mapped to a fixed set of symbolic feature values in the feature space. To recognize symbolic feature states in each feature space, the following clustering scheme is applied to a sample domain with a video database.

A set of N randomly selected videos are picked from the video database for clustering. Every video is randomly subsampled to get S samples. These samples could be single frames or a group of frames (GoFs), each consisting of a group of c continuous frames; the value of c depends on video size. Different low-level media features as required by the media classifiers are extracted for each GoF. These feature values are then clustered using *K-Means clustering* to form K clusters in each feature space. Therefore, for each feature space, $N \times S$ feature values are found that are clustered to get K clusters. The K cluster center values represent K symbolic feature states or media feature "terms" that are available in the dataset. Each other video in the collection is similarly preprocessed and subsampled to extract media features for S GoFs in the video. By performing feature-specific similarity computations of feature values with feature terms, the system can recognize the occurrence of these terms in a video. This media-feature extraction, clustering, and modeling scheme is explained in detail in [129]. Thus computation of probability $P(f_i, f_j)$ is mapped to computation of $P(c_k, d_l)$, where c_k and d_l denote cluster center values that f_i and f_j map to, in their respective feature spaces. After computing mutual information between features, the FBN algorithms for structure and parameter learning can be applied to learn the association of the concept with each feature as well as dependencies between features.

7.5   Application of Multimedia Ontology Learning

In this section, we illustrate the application of the ontology learning scheme in building a sample dance ontology with a digital collection of dance videos. The example dance domain is discussed in detail in chapter 10. We detail each step in the construction process, showing how the ontology is constructed from domain knowledge, then fine-tuned with the help of FBN learning using labeled dance videos. Performance of the ontology learning scheme is discussed, with results of the experiments conducted to measure similarity in structure between the FBN learned ontology and an *expected* ontology as provided by the experts. Results of another set of experiments that validate the parametric learning are discussed. These experiments attempt to *recognize* the various abstract domain concepts with the help of the learned ontology.

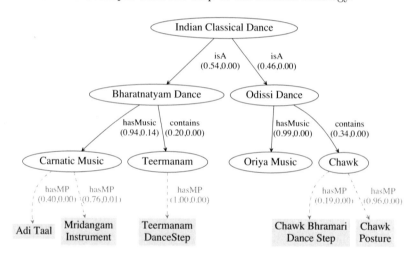

FIGURE 7.5: Basic dance ontology snippet Γ_B as specified by the domain experts, enriched with multimedia.

The basic dance ontology is constructed from specialized domain knowledge gathered from the domain experts and other sources. The ontology is expressed in MOWL. The dance experts also provide their observations on a set of videos, specifying dance-forms, music forms, dance postures, dance steps, hand gestures, name of a dancer, musicians, and similar attributes. that were part of a dance performance. Based on the expert observations, video frames showing dance postures, short video clips containing dance actions, audio files for music forms, and so on, were extracted from the training set of videos. These media files were attached as multimedia examples to the relevant domain concepts in the dance ontology and were also used as training data to train media pattern detectors.

The ontology learning that happens in the ontology learning framework has two aspects:

- *learning the structure* of the ontology, which involves addition and deletion of links in the ontology, thus changing the causal dependencies between concepts, and between concepts and media nodes.

- *learning the parameters* which are the conditional probabilities of the causal relations in the ontology.

The next two sections illustrate the learning of the *structure* and the *parameters* of the sample multimedia ontology. The *parameters* are simultaneously learned in the FBN algorithm and are verified with demonstration of concept-recognition and semantic annotation generated as its consequence.

To illustrate the ontology learning, we take a simple example snippet Γ_B from a basic dance ontology, shown in figure 7.5. This snippet, enriched with media features (leaf elliptical nodes), shows **Bharatnatayam Dance** and **Odissi Dance** as two styles of **Indian Classical Dance**. **Bharatnatyam Dance** is related to the music form **Carnatic Music** and a concept **Teermanam**, which is a dance step typically contained in **Bharatnatyam Dance** performances. Media manifestations of **Carnatic Music** include a musical beat called **Adi Taal** and an instrument **Mridangam Instrument** which is regularly played as part of a Carnatic music performance. The concepts related to **Odissi Dance** are the music accompanying its performances, which is **Oriya Music**, and the concept **Chawk**, which has media manifestations in the form of a posture, **Chawk Posture**, and a dance step, **Chawk Bhramari Dance Step**. MOWL encoding of the ontology is done to associate the expected media patterns with concepts as well as to associate probability values to the CPTs. Some of the probability values come from the domain experts' perspective, while the others are obtained from the training set of videos. The pair of values at each link in the ontology denote the conditional probabilities $P(M \mid C)$ and $P(M \mid \neg C)$, where C is a concept and M represents an associated concept or media pattern.

7.5.1 Learning the Structure

To conduct the experiments for validating the learning of ontology, we need an *expected* version of the ontology as a benchmark, with which we can compare the learned ontology and verify that the structure learned is valid given a bounded error margin. The starting point in the ontology learning process, is the basic ontology Γ_B, constructed from domain knowledge obtained from dance teachers and masters, and enriched with multimedia data from the labeled examples. This ontology represents the domain experts' perspective and encodes the complexities of the background knowledge of the dance domain.

We obtained an expected version of the ontology Γ_E shown in figure 7.6 from a different set of domain experts: the dancers who have contributed

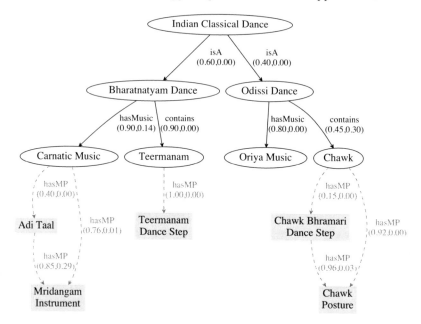

FIGURE 7.6: Expected dance ontology snippet Γ_E as specified by a different set of domain experts.

their dance videos to the digital collection. Their versions of the ontology differ in structure from Γ_B, as the domain concepts and their relationships (as interpreted by the dancers) are more in tune with the current context in which the dance performances take place. The dancers may not have the liberty of adapting the rules and grammar of the classical domain beyond a certain permitted boundary, but they do understand the practical dependencies and co-relations between dance, music, postures, themes and roles in the existing scenario *better* than the theoretical knowledge that the dance teachers might possess. The main difference between the specifications in Γ_B and Γ_E are as follows:

– Chawk Bhramari Dance Step contains the posture Chawk Posture.

– Adi Taal musical beat has Mridangam Instrument as its media observable, as this musical instrument is often used in Bharatnatyam dance performances to play the beat.

Some of the probabilities specified in the two ontologies are also different, but we are not looking at parametric similarity here.

We perform FBN learning on observation models obtained from basic ontology Γ_B, then update the ontology with that learning, to obtain the learned ontology Γ_L. A *graph matching* performance measure is applied to measure the similarity between the two versions : the learnt ontology Γ_L and the expected ontology Γ_E.

Algorithm 6 Applying FBN learning to learn an ontology.

Inputs: a) Basic Ontology Γ_B
        b) Case Data obtained from the training Set of Multimedia Documents
Output: Learnt Ontology Γ_L

1: **procedure** MAIN
2:    Compute $\mathcal{O} = \{ \Omega: \Omega$ is an OM for a concept in $\Gamma_B\}$
3:    **for** $i = 1$ to $|\mathcal{O}|$ **do**
4:        Compute $\mathcal{S} = \{$ s: s is a subnet of height $= 1$ in Ω_i $\}$
5:        **for** $j = 1$ to $|\mathcal{S}|$ **do**
6:            Apply FBN learning with case data to learn s_j
7:            Update Ω_i with s_j
8:        **end for**
9:    **end for**
10:    $\Gamma_L = \Gamma_B$
11:    **for** $i = 1$ to $|\mathcal{O}|$ **do**
12:        Update Γ_L with Ω_i
13:    **end for**
14:    Return Γ_L
15: **end procedure**

7.5.1.1 Performance Measure

A MOWL ontology is a directed, labeled graph, and so are the two versions of the ontology, Γ_L and Γ_E, which need to be measured for similarity. There are several standard similarity measures defined to compute similarity between directed, labeled graphs, which are graphs with a finite number of nodes, or vertices, and a finite number of directed edges. We have chosen *graph edit distance* based on *maximum common subgraph* reviewed in [26]. A maximum common subgraph of two graphs, g and g', is a graph g'' that is a subgraph of both g and g' and with the maximum number of nodes, from among all possible subgraphs of g and g'. The maximum common subgraph of two graphs need not be unique. The larger number of nodes in the *maximum common subgraph* of two graphs, the greater is their similarity.

The other performance measure, *graph edit distance*, provides more error-tolerant graph matching. A graph edit operation is typically a deletion, insertion, or substitution (i.e., label change), and can be applied to nodes as well as to edges. The edit distance of two graphs, g and g', is defined as the "shortest sequence of edit operations" that transform g into g'. Obviously, the shorter this sequence is, the more similar the two graphs are. Thus edit distance is suitable to measure the similarity of graphs. According to [26], the maximum common subgraph g'' of two graphs g and g' and their edit distance $d(g, g')$, are related to each other through the simple equation:

$$d(g, g') = |g| + |g'| - 2|g''| \tag{7.1}$$

where $|g|$, $|g'|$ and $|g''|$ denote the number of nodes of g, g' and g'' respectively.

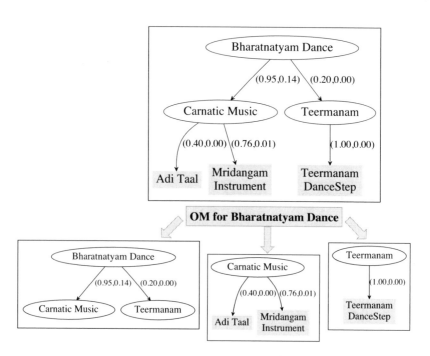

FIGURE 7.7: Observation model of concept `BharatnatyamDance` from Γ_B, split into its subnets for FBN learning.

7.5.1.2 Logic and Implementation

The process of applying ontology learning in terms of obtaining the OMs from ontology, learning the OMs and then updating the ontology with the changed structure and parameters is detailed in algorithm 6. The two inputs to this algorithm are the basic ontology Γ_B and case data obtained from the labeled set of files from the multimedia collection – in this case, the labeled videos from the digital collection. As mentioned in section 7.4.2, a set of naive Bayesian subnets is obtained for each OM extracted from Γ_B. For example, Figure 7.7 shows how the OM for concept node `Bharatnataym Dance` is split into naive Bayesian subnets. `Bharatnataym Dance`, the root node of the OM, is associated with its related concept `Carnatic Music`, which has media patterns `Adi Taal` and `Mridangam Instrument` attached. The CPTs are copied into each subnet from Γ_B. FBN learning from case data takes place in each Bayesian subnet, updating its structure and parameters. Each learned subnet is then used to update its parent OM. All such learned OMs are then used to update the ontology. The output of the algorithm is the learned ontology Γ_L.

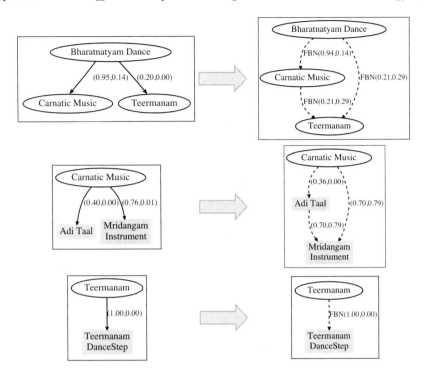

FIGURE 7.8: Subnets of `BharatnatyamDance` OM updated with FBN learning.

Figures 7.8 and 7.9 show the FBN learning and subsequent updation of the OM for concept `Bharatnatyam Dance`. Equation 7.3 is used to compute $W(\texttt{Bharatnatyam Dance}) > W(\texttt{Carnatic Music}) > W(\texttt{Teermanam})$, for concepts in `Bharatnatyam Dance` subnet in figure 7.8. Accordingly, an ordering is imposed on the nodes in the subnet to generate a new structure. Once CPTs are learned, as detailed in the next section, the parameters determine whether all the links are retained or some are deleted. The `Carnatic Music` subnet in figure 7.8 illustrates the association of the concept `Carnatic Music` with its media manifestations `Adi Taal` and `Mridangam Instrument`, along with the new ordering amongst the media nodes, learned through the FBN technique. Similar to the learning of `Bharatnatyam Dance` subnet, the splitting of the OM for `Odissi Dance`, its FBN learning and updation is shown in figures 7.10 and 7.11.

Applying the graph similarity performance measures, we first find the *maximum common subgraph* Γ_C of the two graphs Γ_E and Γ_L. Then *graph edit distance* between the two graphs is computed as follows:

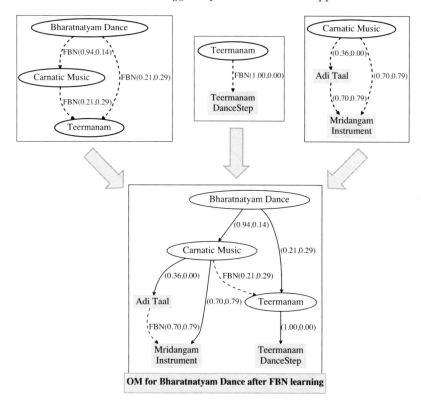

FIGURE 7.9: Observation model of concept `BharatnatyamDance` updated after FBN learning.

$$d(\Gamma_E, \Gamma_L) = |\Gamma_E| + |\Gamma_L| - 2|\Gamma_C| \qquad (7.2)$$

where $|\Gamma_E|$, $|\Gamma_L|$ and $|\Gamma_C|$ denote the number of nodes of Γ_E, Γ_L and Γ_C, respectively. As we can see here, the $d(\Gamma_E, \Gamma_L) = 2$ for the example snippet ontology shown here. Around 30 percent concepts in the dance ontology were at a suitably high abstract level for their observation models to be tuned with the FBN learning algorithm. Experiments were done with a large number of observation models, with the number of nodes in the OMs ranging from 6 to 10 and the number of edges ranging from 5 to 10. An average performance of *graph edit distance* $= 2.4$ between the learned and expected versions was obtained.

7.5.2 Parametric Learning

FBN learning from case data leads to a change in the structure and parameters of the Bayesian network representing the OM extracted from the

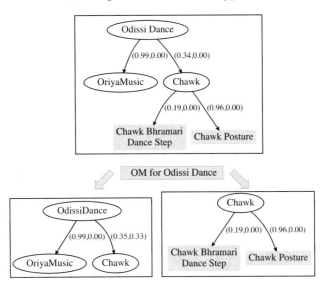

FIGURE 7.10: Observation model of concept Odissi Dance from Γ_B, split into its subnets for FBN learning.

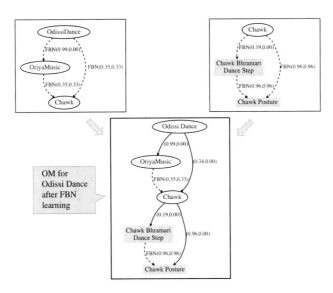

FIGURE 7.11: Observation model of OdissiDance, with its FBN learning and updation.

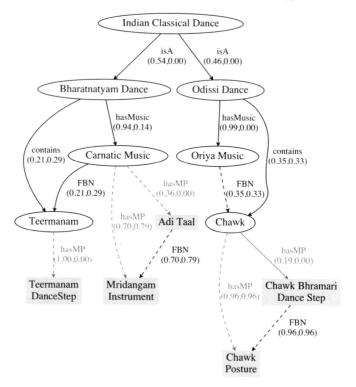

FIGURE 7.12: Learned dance ontology Γ_L with its structure and parameters changed due to FBN learning from case-data.

ontology. After all the OMs have been learned, they are used to update the structure of the ontology and also to change the joint probabilities encoded in the ontology according to the new parameters learned in the OMs. The ontology learned in this manner is dynamic, as it can be refined and fine-tuned automatically with additions to the video database. The newly learned ontology can then be applied afresh to recognize concepts in the video database. If the learning is good, then the concept detection and subsequent annotation generation should show improved results with the fine-tuned ontology. We discuss a small example to demonstrate how concept detection improves with the domain ontology changed after applying FBN learning.

7.5.2.1   Concept Recognition Using MOWL

Once an OM for a semantic concept is generated from a MOWL ontology, the presence of expected media patterns can be detected in a digital multimedia artifact using appropriate media detector tools. Such *observations* lead to instantiation of some of the media nodes in the OM, which in turn results in belief propagation in the Bayesian network. The posterior probability of the

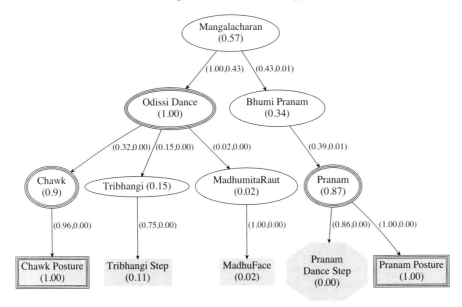

FIGURE 7.13: Concept recognition in the `Mangalacharan` OM generated from the basic ontology Γ_B.

concept node as a result of such belief propagation represents the degree of belief in the presence of the semantic concept in the multimedia artifact.

For example, figure 7.13 shows the BN representing the OM of concept `Mangalacharan` – some media patterns have been detected in an Odissi dance video and corresponding media nodes in the OM have been instantiated. The links between concept nodes, between media nodes and between a concept and a media node denote causal relations as well as uncertainty specifications that have been learned from data. A bracketed value with the name of each node denotes its posterior probability after media nodes have been instantiated and belief propagation has taken place in the BN. In this video, the media patterns detected with the help of concept detectors are `Chawk Posture` and `Pranam Posture`, shown as gray ellipses. If the posterior probability of a concept in the OM is above a threshold value (0.65), there is sufficient belief to conclude its presence in the dance video, then it is deemed to be *recognized*. The concept nodes `Chawk` and `Pranam` highlighted in gray boxes are the low-level concepts that are recognized due to presence of these media patterns in data. Due to belief propagation in the BN, a higher-level concept, `Odissi Dance`, is recognized (colored in gray) to be present in the video.

7.5.2.2 Concept Recognition after Learning

Figure 7.14 shows the OM of the concept `Mangalacharan` after FBN learning has been applied to it. The OM is constructed from the dance ontology

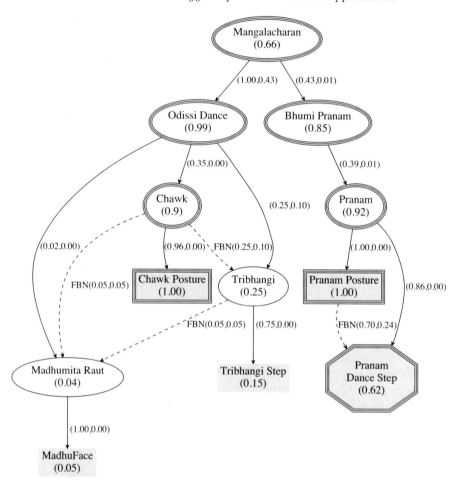

FIGURE 7.14: Bayesian network corresponding to the observation model of the concept Mangalacharan after FBN learning has caused the structure and parameters to be updated.

refined with FBN learning, so the probability values shown correspond to real-world data. After applying FBN learning, some new relations (shown with *dotted* links and labeled FBN) were added, based on statistical evidence in case data.

Let the media patterns detected in the test Odissi dance video be Chawk Posture and Pranam Posture. The corresponding media nodes are instantiated in the Mangalacharan OM generated from the FBN learned dance ontology. As in the earlier case, Chawk and Pranam are the low-level concepts that are recognized due to the presence of these media patterns in data. Due to an FBN link between `PranamPosture` and `PranamDanceStep`, the latter node is also instantiated, thus leading to higher belief in the presence of concept `BhumiPranam` in the video. Higher-level concept nodes (in gray color) are recognized to be present due to belief propagation in the BN. `Chawk` pattern causes `Odissi Dance` to be recognized. Presence of `Pranam` and `BhumiPranam` lead to recognition of `Mangalacharan` concept, which is further confirmed by recognition of `Odissi Dance` concept in the video. This concept recognition is confirmed by the labels that the domain experts had provided for the test `Odissi dance` video. Thus FBN learning has led to an improvement in concept-recognition as in the basic ontology, only one abstract concept `Odissi Dance` was recognised in the video.

Concept	Basic					FBN				
	C*	M*	F*	P*	R*	C*	M*	F*	P*	R*
Adi Taal	52	12	20	0.72	0.81	54	18	12	0.82	0.75
Battu Dance	15	3	12	0.55	0.83	22	4	4	0.85	0.85
Carnatic Music	92	10	3	0.97	0.90	97	5	3	0.97	0.95
Group Dance	12	13	9	0.57	0.48	27	3	4	0.87	0.90
Krishna Role	53	35	10	0.84	0.60	73	10	15	0.83	0.88
Krishna Sakhi Theme	1	2	3	0.25	0.33	2	1	3	0.40	0.67
Madhumita Raut	11	10	12	0.48	0.52	22	2	9	0.71	0.92
Mahabharat Theme	7	15	3	0.70	0.32	15	3	7	0.68	0.83
Mangalacharan Dance	23	36	13	0.64	0.39	53	14	5	0.91	0.79
Naughty Krishna	6	3	5	0.54	0.67	5	4	5	0.50	0.55
Pranam	43	10	23	0.65	0.81	55	8	13	0.80	0.87
Solo Dance	26	13	18	0.59	0.66	41	8	8	0.84	0.84
Vanshika Chawla	22	5	3	0.88	0.81	27	2	1	0.96	0.93
Yashoda Role	7	2	1	0.87	0.77	5	3	2	0.71	0.63

TABLE 7.1: Table showing high-level annotation results using basic and FBN learned ontology. Legend: C=Correct, M=Miss, F=False, P=Precision, R=Recall.

7.5.2.3   Semantic Annotation Generation

An important contribution of this ontology learning scheme is the attachment of conceptual annotation to multimedia data belonging to a domain, thus preserving its background knowledge and enhancing the usability of this data through digital access. Figure 7.15 shows the architecture of an annotation generation framework. It consists of five functional components. The basis of this whole framework is the MOWL ontology created from domain knowledge, enriched with multimedia data, and then refined with learning from annotated examples of the domain. The most important component of this process is the *Concept-Recognizer*. The task of this module is to recognize the high-level semantic concepts in multimedia data with the help of low-level media-based features. OMs for the high-level concepts selected by the curator of the collection are generated from the MOWL ontology by the *MOWL*

Parser and given as input to this module. Low-level media features (SIFT features, Spatio-temporal interest points, MFCC features, etc.) are extracted from the digital artifacts, which can be in different formats (image, audio, video), and provided to the *Concept Recognizer* by *Media Feature Extractor*. *Media Pattern Classifiers*, trained by feature vectors extracted from the training set of multimedia data, help detect the media patterns (color, shapes, texture, objects, postures, audio, etc.) in the digital artifacts. In initial stages of building the ontology, data is labeled with the help of *manual annotations*, provided by the domain experts in XML format.

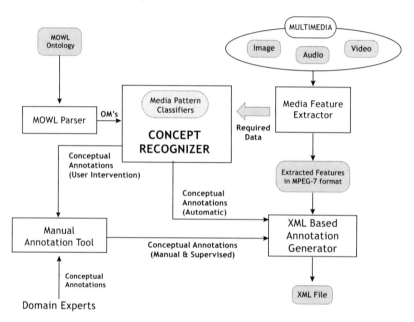

FIGURE 7.15: Architecture of annotation generation framework

To recognize concepts in a test multimedia document, evidence is gathered at the leaf nodes, as different media features are recognized or classified in the document by the media classifiers. If evidence at a node is above a threshold, the media feature node is instantiated. These instantiations result in belief propagation in the Bayesian network, and posterior probability at the associated concept nodes is computed. After belief propagation, these nodes have high posterior probability. As they get instantiated, we find high belief for the existence of other high-level abstract nodes. Conceptual annotations are generated and attached to the video through Semantic annotation generation. Results for some of the conceptual annotations generated for the dance domain using the basic ontology and then the FBN learned ontology are shown in table 7.1. On average, we see an improvement in *Precision* and *Recall* from the FBN learned ontology over the results from the basic ontology.

7.6 Conclusions

In this chapter, we discussed learning in an ontology from data, noting that while existing techniques work well for ontologies that are constructed from textual data, there is a gap when it comes to learning a multimedia ontology from media. Learning an ontology is a complex task, which gets more complicated when the dataset is media based and the ontology to be learned has to encode multimedia-specific properties and relations, as well as compute uncertainties linked with media observables. In this context, we discuss a technique by which the knowledge obtained to construct an ontology from a domain expert can be fine-tuned by applying learning from real-world multimedia examples belonging to the domain. An ontology refined in this manner is a better structured, more realistic model of the domain that it represents. We also discussed a novel technique to populate a collection of videos from a knowledge domain in order to provide a flexible, ontology-driven access to the users of the domain. The ontology learned from video examples represents a domain more attuned to real-world data. The MOWL ontology learned in this manner not only has a more refined *structure* as is proved by experiments, but also has more accurate conditional *probabilities* encoded in the CPTs attached to its concepts and media-based nodes.

Chapter 8

Applications Exploiting Multimedia Semantics

8.1 Introduction

As we have seen in the previous chapters, there are many aspects to multimedia semantics. Multimedia contents may represent various objects or events, some of which may have contextual importance. Some obvious examples are discovery of a firearm in the x-ray image of baggages scanned at an airport or interpretation of a sequence of frames in a video as a sports event, such as scoring of a goal in a soccer tournament. There are other subtle aspects of multimedia semantics beyond object or event recognition. For example, different configurations of the human faces and different composition of melodies depict various human emotions, and the brushstrokes on paintings often characterize the artists or the schools they belong to.

In this chapter, we deal with the exploitation of deep semantics of multimedia artifacts in some real-life applications. We have dealt with three major classes of applications – information retrieval, content recommendation, and information integration – which are of general interest. We show that how the multimedia knowledge representation scheme proposed in chapter 5 of this book enables extracting high-level semantic attributes of multimedia artifacts and space-time localization of multimedia events in different contexts. We start with the problem of retrieval from heterogeneous media repositories in the complex domain of Indian culture and its extension to include document images. Then we move on to recommendation problems for media-rich artifacts, such as paintings and garments, where there can be a good deal of subjectivity in contextual interpretation and aesthetics. Next, we provide examples of news aggregation from multiple sources, including professional websites as well as social media, in different media forms. Finally, we deal with event localization in space and time with an extension of MOWL.

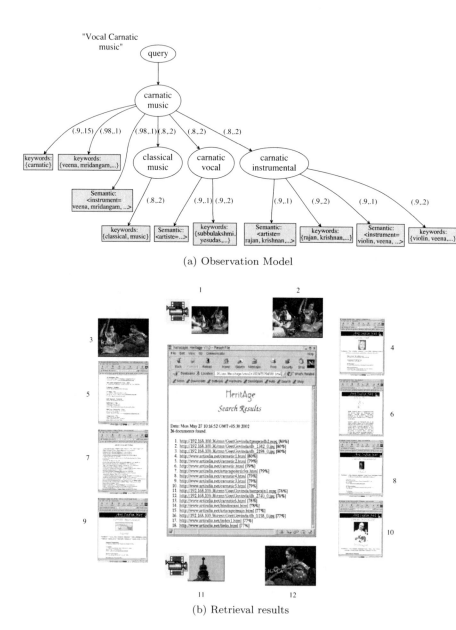

(a) Observation Model

(b) Retrieval results

FIGURE 8.1: Retrieval for the query "vocal Carnatic music" in different media forms.

8.2 Multimedia Retrieval and Classification

Though there are content-based retrieval engines for different media forms like images, music and video, there are no retrieval engines that can interpret media in different forms for content-based retrieval. The difficulty in semantic retrieval from multiple and heterogeneous media repositories arises out of the facts (a) that different methods need to be applied to interpret the contents in different media forms, and (b) that there is no basis for harmonizing the retrieval scores (estimated probability for relevance) obtained through application of different methods in the different media collections. Use of a common domain knowledge that spans across multiple media modalities can alleviate this problem. The evidential model of reasoning associated with MOWL allows us to exploit the redundancy in the observation model and to create customized observation plans for individual media collections. At the same time, since the observation plans are derived from a common knowledge base, the relevance scores produced by the different methods at the different collections have a common basis for harmonizing the relevance scores and hence ranking of the combined result set.

8.2.1 HeritAge: Integrating Diverse Media Contents from Distributed Collections

HeritAge [65] has been conceived as a virtual encyclopedia for Indian heritage. It uses multimedia knowledge representation for semantic retrieval of information from a number of repositories on the web, hosting different media forms – namely text, images and videos, – in the domain of Indian cultural heritage. Figure 8.1 shows the observation model and the retrieval results for a query "vocal Carnatic music" (classical music from the southern states of India). Note that the results in different media forms (text, image and video) have been obtained from different collections on the Internet.

8.2.2 Document Image Classification

Document images of various kinds, such as ancient manuscripts, old books, research papers, newspapers and handwritten medical prescriptions account for a significant portion of the contents in the current digital collections, motivating their indexing and retrieval. The identity of a document image often manifests into its structural components and their relative layout. For example, a newspaper front page is identified by its bold title, the news headlines and the column-oriented structure. Figure 8.2 shows sample document images to illustrate the point. Thus, the hierarchical structure of a document, identity of the structural elements and the layout of the page (i.e. relative placement of the structural components) constitute important cues for classification of document images.

(a) Newspaper

(b) Content page of a book

MOWL: An Ontology Representation Language for Web based Multimedia Applications

ANUPAMA MALLIK, Indian Institute of Technology, Delhi
HIRANMAY GHOSH, Tata consultancy Services
GAURAV HARIT, Indian Institute of Technology, Jodhpur
SANTANU CHAUDHURY, Indian Institute of Technology, Delhi

Several multimedia applications need to reason with concepts and their media properties in specific domain contexts. Media properties of concepts exhibit some unique characteristics that cannot be dealt with conceptual modeling schemes followed in the existing ontology representation and reasoning schemes. We have proposed a new perceptual modeling technique for reasoning with media properties observed in multimedia instances and the latent concepts. Our knowledge representation scheme uses a causal model of the world whose concepts manifest in media properties with uncertainties. We introduce a probabilistic reasoning scheme for belief propagation across domain concepts through observation of media properties. In order to support this perceptual modeling and reasoning paradigm, we propose a new ontology language, Multimedia Web Ontology Language (MOWL). Our primary contribution in this paper is to establish the need for the new ontology language and to introduce the semantics of its novel language constructs. We establish the generality of our approach with two disparate knowledge-intensive applications involving reasoning with media properties of concepts.

Categories and Subject Descriptors: I.2.4 [Computing Methodologies]: ARTIFICIAL INTELLIGENCE—*Knowledge Representation Formalisms and Methods*

General Terms: Theory,Standardization

Additional Key Words and Phrases: Multimedia Web Ontology Language, OWL, Ontology Bayesian networks, Video Annotation

ACM Reference Format:
Mallik, A. and Ghosh, H. and Harit G. and Chaudhury, S. 2011. MOWL - Representation Language for a Multimedia Ontology ACM Trans. Multimedia Comput. Commun. Appl. 2, 3, Article 1 (November 2011), 28 pages.
DOI = 10.1145/0000000.0000000 http://doi.acm.org/10.1145/0000000.0000000

1   INTRODUCTION

A large part of multimedia data in various web-based repositories and social networks remains to be adequately annotated. Such data need to be semantically interpreted in various application contexts. Domain knowledge plays a crucial role in semantic data processing in knowledge-intensive domains.

Authors' address: A. Mallik, Multimedia Lab, Block D, IIT Delhi, New Delhi-110016, India; email: anumallik@gmail.com; H. Ghosh, TCS Innovation Labs Delhi, 249 D&E DLF City Phase IV, Gurgaon-122002, India; email: hiranmay.ghosh@tcs.com; G. Harit, IIT Jodhpur, Jodhpur; email gharit@iitj.ac.in; S. Chaudhury, Multimedia Lab, Block D, IIT Delhi, New Delhi-110016, India; email schaudhury@ee.iitd.ac.in;

(c) Research paper

FIGURE 8.2: Illustrative document layouts for discrimination.

While the layout characterizes the nature of a document, the semantics of its contents manifests in the text, which appear inthe different structural components. There are many situations where OCR technologies cannot be reliably applied on these document segments to extract the text. The reasons include non-availability of reliable OCRs in some ancient or regional languages and the quality of document scans. An interesting approach to overcome the problem is to recognize the words, which are the atomic entities carrying semantics [106, 85]. While different authors may have used different image features, the general principle behind this approach is that a larger feature set for a word results in more robust recognition than a single letter.

The new multimedia ontology representation scheme presented in this book can effectively encode the structural components of a document in terms of their media features, spatial layouts and hierarchy. It can also encode the spatial organization of the features (usually, the shape primitives) that represent a word. The use of multimedia ontology in classification of documents based on its hierarchical structure and word-based indexing has been proposed in [84]. As a specific example, sections of representation for a newspaper layout are illustrated in figure 8.3.

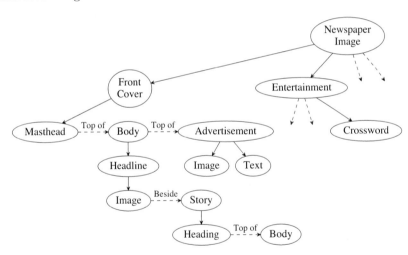

FIGURE 8.3: Encoding of newspaper layout.

This approach for interpreting document images had been applied to extend the virtual encyclopedia of Indian cultural heritage discussed in the previous section. This extension resulted in the inclusion of document images over and above other heritage artifacts in the encyclopedia. Semantic relations have been established across the different types of media artifacts using the MOWL ontology, facilitating an integrated view of the repository contents.

8.3   Recommendation of Media-Rich Commodities

Retail shopping portals like Amazon, Flipkart and eBay have gained enormous popularity over the past few years for providing a convenient shopping platform for a wide range of products. There are other portals selling specialized products, such as paintings, interior decoration items, high-end fashion accessories and garments. The selection of the items in such portals is primarily guided by their subjective aesthetic value and compatibility with the context of usage. For example, choice of fashion accessories are generally guided by the wearer's physical attributes, the occasion of use and wearer's personal sense of aesthetics. Conventional catalogs can, at best, classify the products based on their genre, price-range, and such attributes. They provide little help in selection of such "media-rich" products since the users need to inspect and assess each of the products for its perceptual properties. Further, the recommendation engines commonly available on shopping portals use collaborative filtering techniques that do not analyze the product properties and are of little help for selecting media-rich products. Successful recommendation for such products requires that the suitability of the products be ascertained by interpretation of their perceptual properties, – for example, the brush-strokes of a painting or the texture of a fabric – in the domain context. This represents a new content-based filtering approach to recommendation, where the contextual semantics of the media features of the products are established with the help of domain knowledge. In this section, we review two implementations for such recommendation engines, one for paintings and the other for garments.

8.3.1   Painting Recommendation

An intelligent painting recommendation system interprets an abstract user request in terms of the different concepts in the domain. The concepts include genre of the painting, associated movements, artists, and place and time in history, to name a few. For example, a user's request for "cubism" may be satisfied by recommending the works of Pablo Picasso, Georges Braque and their followers. As another example, a user interested in "folk art from India" can be satisfied with Kalamkari, Madhubani and such other genres of paintings that are linked to various geographical part of India. Each of such paintings are also associated with specific media patterns. For example, "cubism" is characterized by straight edges, a "van Gogh" by some characteristic brush-strokes and "Madhubani" by its vibrant colors. Figure 8.4 depicts examples of such paintings. Knowledge about such media properties over and above other domain relations can be encoded in MOWL.

Windowshopper [66] is a painting recommendation system that exploits knowledge about media properties of the various concepts in painting domain. It exploits an agent-based architecture (discussed in detail in chapter 9) to

| (a) Cubism | (b) van Gogh | (c) Madhubani |

FIGURE 8.4: Different painting styles (images are sourced from Wikimedia Commons).

integrate several painting retail portals in a common ontological framework. The reasoning model in the recommendation system follows that discussed in chapter 5.

FIGURE 8.5: Recommendations from Windowshopper: a watercolor similar to a van Gogh painting shown at the right.

8.3.2   Garment Recommendation

Shopping for a garment for a specific occasion is a daunting task. A garment needs to have an aesthetic appeal on the body of the wearer and should be befitting to the occasion. Assessing contextual suitability of the garments requires visual and tactile examination of each product and poses a significant cognitive load on the buyer. The problem can be alleviated with a recommendation engine that can analyze the *look and feel* (visual and tactile properties) of the garments in light of the knowledge about the fashion domain and in the context of the wearer as well as the occasion. An ontology-based garment recommendation engine has been proposed in [199], where human body and garment attributes have been organized in taxonomies and fashion experts' views on the suitability of garment attributes with respect to human body attributes have been encoded as recommendation rules. For example, a "tight-fit" jacket is recommended for "normal" body shape and "warm" colors are recommended for a person with "spring" color season [111]. Moreover, there are some rules to account for social occasions like a "dinner party". This recommendation system may be classified as a content-based recommendation system, since the recommendation rules are based on interpretation of content properties and not based on the purchase history of the user groups.

It may be noted that the "rules" in the fashion domain are largely subjective choices and can vary across individuals as well as experts. Thus, a fashion recommendation engine should not be prescriptive but should help a user in making his or her choice by narrowing the search space. The use of the crisp rules in [199] does not account for personal preferences and makes the system restrictive. Moreover, the fine-grained media properties of the garments that constitute important selection criteria are not analyzed in this approach. An alternate approach based on statistical analysis of past wearing habits of the user has been proposed in [123]. This approach attempts to acquire the knowledge about an individual's fashion preferences disregarding any expert opinion. It attempts to provide personalized recommendations in accordance with the individual's choice from his or her personal wardrobe.

An apparel recommendation system based on perceptual modeling of concepts with MOWL has been presented in [1]. The system deals with "Sarees", an ethnic wear for women in the Indian subcontinent. The choice of a Saree befitting an occasion is a highly subjective matter and depends on several factors, such as the wearer's physical attributes, socio-cultural background and demography. The ontology representation scheme of MOWL provides an opportunity to encode the media properties of the fashion concepts with uncertainties and to establish soft reasoning rules to realize a flexible garment recommendation system. A section of the fashion ontology that has been used in the system is shown in figure 8.6. It incorporates knowledge about human users, occasion to wear, the garment properties and their contextual associations. The evidential model of reasoning supported with MOWL allows for

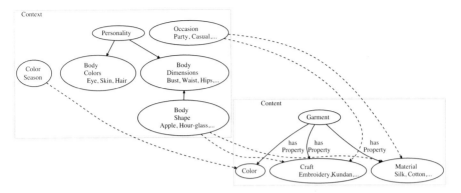

FIGURE 8.6: Ontology for garment recommendation.

building in redundancy in the system and can produce robust results despite variations and incompleteness of product catalogs.

A woman is said to be "well-dressed" when the visual properties of her dress blends well with her personality and the occasion. Fashion experts' views on such "blending" are encoded as the causal associations between the fashion concepts and control the media property flow, rather than a rule base as in [199]. For example, the body colors of a human as portrayed in her "color season" affect the suitability of the color properties of the garments. The relationship is captured with a soft link between the two entities. Similarly, body shape of a person impacts the choice of material and craft of the garment. The domain knowledge is derived from Color Season Model [111] and many other information sources available on the web. The informal qualitative information available at these sources has been quantified to compute the conditional probabilities in the ontology using several numeric and media data-analysis techniques. For example, a probabilistic model for color season model has been computed with Gaussian Mixture Model (GMM) based analysis of the color values obtained from the color charts, such as in [154]. Though important, the socio-cultural background of the wearer and demographic information is difficult to ascertain and have not been incorporated in the system.

The recommendation problem is handled in two steps in the system. In the first stage, a visual profile of the user is determined. It consists of two parameters, namely the color season (summer, spring, etc.) and the body shape (pear, hourglass, etc.) [217]. The problem is modeled as that of evidential reasoning. The color season of a user is believed to cause some visible body colors (skin, hair and eye colors) and the body shape is believed to result in some measurable body dimensions (shoulder, bust, waist and hip measurements). The continuous range of colors and body dimensions are quantized into finite number of states. Two naive Bayesian networks as shown in figure 8.7 are deployed to determine the visual profile with the evidence of visible body colors (determined from the face images of the user [44]) and declared body

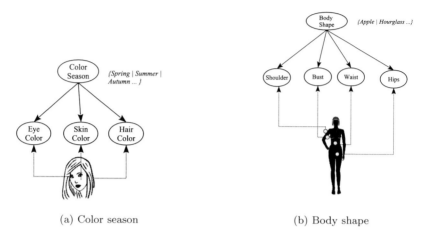

(a) Color season   (b) Body shape

FIGURE 8.7: Observation models for visual profile.

measurements. The knowledge about such dependencies have been *apriori* encoded in the ontology and the Bayesian networks (the observation models for the visual profile) have been drived by reasoning with it.

In the second stage of reasoning, an observation model for the "suitable" garments, as shown in figure 8.8 is created. A suitable garment manifests itself in some garment properties, some of which relate to user visual personality and some to the occasion of wear (with possible overlap). The intermediate nodes in the observation model are accordingly labeled. The leaf nodes in this OM represent "observable" garment properties. The suitability of each of the garments available in the store are analyzed with this OM. The properties of the individual garments in the stores are obtained either from the associated metadata or by analyzing the garment image available in the catalog. The leaf nodes on the OM are instantiated based on these observations. The posterior probability of the root node on instantiating the property values observed in a garment is an estimate of its suitability in the given context. Finally, a ranked list of garments is computed for recommendation, where the rank of a garment is determined by its relevance (suitability) as well as its diversity from the garments already in the list [230]. It may be observed that the garment metadata may be incomplete, that is, all the attributes used in the OM may not be available in the catalog. The evidential reasoning model is robust again such partial availability of data.

Validation of a garment recommendation system is a difficult task, because of inherent subjectivity associated with the fashion domain and the unavailability of labeled data that can be treated as ground truth. One way to validate the system is with subjective user ratings. The MOWL-based garment recom-

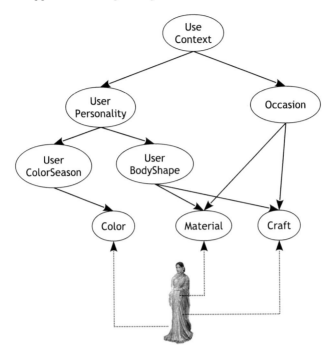

FIGURE 8.8: Observation models for "suitable" garment.

mendation engine achieved an encouraging median user satisfaction rating of 3.7 on a 4-point scale for party-wear. Another way to validate the recommendation engine is to compare the recommendation results for celebrities with the garments actually worn by them (assuming that the celebrities or their wardrobe managers have a good sense of fashion). The Sarees actually worn by a group of Indian celebrities were found to tally significantly more with the recommendations rather than random pick-ups from the collection. KL-divergence of the GMM's of the Saree attributes formed the basis of such comparison [87].

8.3.3 Discussions

In this section, we have presented a novel approach to recommendation for feature-rich products. Though this method can be broadly classified as content-based filtering, it is quite distinct from existing approaches. Existing recommendation engines based on content-based filtering [27, 121] use product features to establish similarity between the products and recommend products similar to those procured by the user earlier. In contrast, the approach presented in the section uses domain knowledge to analyze the product features and to establish suitability of the products in a given context. Expressing the domain knowledge with MOWL helps in dealing with media-based features that are important for media-rich products. The soft "rules" connecting the different domain concepts naturally account for subjectivity in perception of aesthetics. The abductive reasoning results in robust inferencing despite incomplete product or context information that can be common to many use-case scenarios. Further, it is possible to incorporate a feedback loop in the system to learn the user's preferences and tune the observation model for recommendation accordingly. Thus, this approach provides an opportunity to combine expert knowledge with personal preferences.

8.4   Information Integration from Open Resources

In a web-based world, information exists in heterogeneous multimedia forms and is fragmented over multiple resources. For example, current news can be acquired from websites hosted by news agencies and TV channels, as well as from social networks. Aligning the information sources and creating a concise multimedia presentation of the information poses a major challenge. We present a few applications that attempt to solve the problem. The common thread that binds these applications is that the integration is based on the content semantics and not on their physical format, source or associated metadata (e.g., keywords) alone.

8.4.1   News Aggregation from Social Media

It has been observed that while authentic and detailed news can be obtained from the websites of established news agencies, nascent news is first broadcast on social media, in particular by microbloggers. The volume of discussion on the microblogger sites is an indicator of the impact factor of a news item. This motivates TWIPIX [10], Me-Digger [11] and Veooz (www.veooz.com) to use Twitter to discover emerging news and present it in multimedia format by aligning the news with the contents from other social networks like Flickr, Twitpic, and Instagram, as well as professional websites hosted by news agencies, like the BBC and the Times of India.

While Bao et al [11] characterized the news with keywords and detected "emerging" news in terms of keyword statistics over finite periods of time, Bansal et al [10] employed natural-language processing and statistical machine-learning techniques to extract the actor, action and the object in the microblogs and characterize an event with a triplet [112]. Several micro-blogs with diverse texts were aggregated to identify an "event". The structured definition of the events facilitated semantic linking of the isolated events discovered from the tweets. The events were chronologically linked to depict stories evolving over time. The URLs associated with the microblogs often pointed to news articles and to images. Analysis of such contents contributed not only to enriching the events and stories with rich multimedia content but also in aligning and aggregating the noisy tweets. While natural-language processing of the short and noisy microblog text proved to be difficult, restricting the system to a specific domain (e.g., London Olympics 2012) yielded acceptable results. It may be argued that use of a domain ontology could help in better resolution of named entities, as well as events. The "interestingness" of an aggregated event was determined by a combination of several parameters, including its information value measured in terms of volume of text and other media artifacts associated with the event as well as its popularity, recentness and novelty. The multifaceted interestingness parameter made popular events give way to nascent events that were yet to be noticed.

8.4.2 Information Aggregation by Event Pattern Detection and Trend Analysis

Today, development of techniques to categorize documents into contextually meaningful classes has become nontrivial with increasing amounts of multimedia information being available. The information available in the documents is composed of a sequence of events that can be termed "patterns". It is interesting to analyze important trends as observed from event patterns that emerge over a specific time period and space. we need to focus on the semantics of the documents for identifying these patterns. A news event encodes a story related to some news topic. These event stories comprise a sequence of events or patterns that occurred at some specific time and space. While searching event-related news, the users might be interested in knowing the major events along with their development temporally as well as spatially. For instance, the user might be interested in knowing trends about how the public protest in Turkey evolved over time and how different parts of the world interacted. Manual tracing such information from different web-based text and video resources is a time-consuming, cumbersome and an almost impossible task. Several research efforts [113, 213, 222] have focused on identifying event patterns using text mining from data and trend analysis.

In this section, we discuss a scheme for identifying the conceptual evolution of event patterns from news videos – temporally as well as spatially – using a multimedia ontology [149]. Once a video is chosen from YouTube or a news-site, and a reasonable transcript of its audio is provided, the application presents temporal and spatial trends along with contextually relevant additional information with the help of an ontology. To achieve this, initially a *Document Set* corresponding to an input video that provides all information available on the web is built and is associated with the video. An event ontology encoded in E-MOWL (an event-based extension of MOWL) [150] provides a semantically plausible way to capture the contextual meaning of a media document. The highly probable words of each topic as obtained using LDA are mapped onto the E-MOWL concepts in a domain-specific ontology to as-

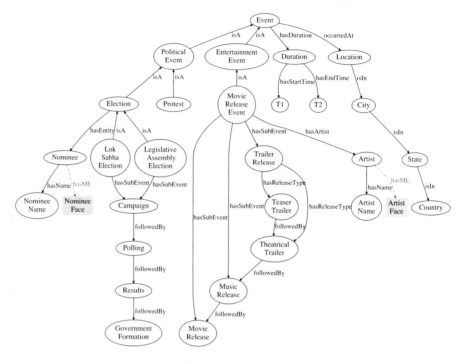

FIGURE 8.9: Event ontology.

sociate a conceptual tag to documents over time. This representation provides a concise structure and insights of occurrence of patterns as a time-ordered stream. For identifying the event patterns geographically, geographic entities are extracted from the documents. The relatedness between the geographic entities from various documents is exploited using a geographical ontology (Geo-ontology), which is also encoded in MOWL.

8.4.2.1   E-MOWL and Geo-ontology

The application leverages an E-MOWL-based ontology and geographic ontology to identify the video-based event patterns over time and space. An E-MOWL-based ontology encapsulates the definitions of events along with relationships that exist between them. The events are encoded with multimedia data, which can be in the form of video, image or audio. They contain a story related to a sequence of events that has already occurred or may take place in future, and other aspects related to it. E-MOWL language constructs are utilized to associate event information in classes, individuals and properties. E-MOWL considers two main classes to represent its basic entities: *Concept* and *EventObject*. The <emowl:Concept> class corresponds to real-world objects or events. <emowl: EventObject> class is composed of following three subclasses: <emowl:SpatialObject> for the location where the event took place; <emowl:TemporalObject> for the day, date and time related information of an event; <emowl:ActorObject>: for actors or entities involved in the event. An example ontology snippet depicting various concepts and their relationships corresponding to two domains (entertainment and politics) is shown in figure 8.9.

The Geo-ontology is concerned with the conceptual specification of geographically meaningful relations between the various geographic entities. The Geo-ontology can formally be defined as $\{C, R, A, O, I\}$, where C (concept) represents the concept set comprising geographic entities; R (relaytion) is a relation set, describing the geographic relations as observed in various geographic entities; A (attribute) shows the various geographic attribute set of objects; O represents the Observation Model of the event in the MOWL probabilistic framework (the entire reasoning is based on this); and I (instances) is a set of definitions about observed media instances. A new abstract class <gmowl:GeoLocation> is defined to specify the geographic properties of observable media objects. Two further subclasses of <gmowl:GeoLocation> are defined as: <gmowl:GeoPhysicalLocation> for physical features of the geographic entity (say, its position, orientation etc.); and <gmowl:GeoPoliticalLocation> for political features of a geographic entity (say, information about state, country etc.).

8.4.2.2 Video Context and Document Set

The context information corresponding to a video is identified by using speech-to-text conversion of its audio content to generate a text document. The document is syntactically and semantically analyzed to identify the contextual information present in that document as shown in Fig. 8.10(a). In syntactic analysis, the text corresponding to a video is analyzed using the Stanford Natural Language Processing (NLP) parser. Using this, the named entities and part-of-speech tags are identified. The named entities (like date and location) help in gaining knowledge about the temporal and geographical aspects of the video. Semantic analysis using E-MOWL-based ontology follows this sysntactic analysis.

The actions or the events present in a video can be represented using verbs. The subject of the verb can be considered as the actor/agent of that subevent. Instead of mapping just only one verb, all its synsets from WordNet [135] are used for mapping onto the subevents present in the E-MOWL-based ontology. The ontology-based event detection involves two steps. In the first step, the OM of the event is constructed. The second step involves reasoning with the OM. The concept nodes corresponding to various subevents as identified are then instantiated. Using belief propagation, the higher-level event is detected. Thus, apart from temporal and geographic aspects, a video has been associated with its event-related information. This is how the contextual tags get associated with a video. Figure 8.10(b) illustrates this framework.

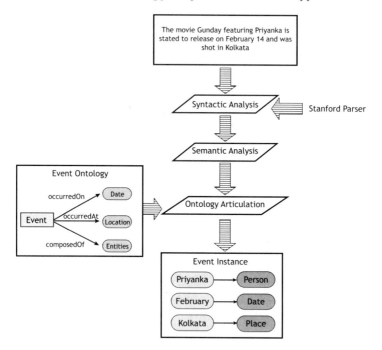

(a) Syntactic and semantic analysis

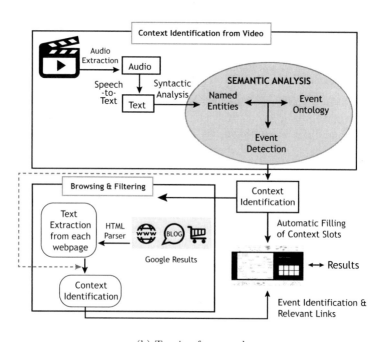

(b) Tagging framework

FIGURE 8.10: Frameworks for video context analysis.

The web is crawled with these contextual tags to get more videos and text news from diverse web-based resources. The retrieved results consist of a mixture of documents. From this, the retrieved videos are then analyzed using the same procedure as followed for initial video (i.e., audio content of the video is converted to text for syntactic and semantic analysis). Now, the corpus of documents consists of transcripts of audio and the text news retrieved after crawling the web. The context of the video is then matched with the syntactically and semantically analyzed context of all the documents. This set of documents (transcripts, text news) associated with a video on the basis of their contextual meaning, forms the *Document Set*.

8.4.2.3 Spatial and Temporal Trend Analysis

The *Document Set* or the *corpus*, comprising videos, transcripts and text news, provides a complete description of the news. This facilitates detecting and correlating trends corresponding to a video. Approximately 100 videos corresponding to two domains – namely *Political* and *Entertainment* – were used for experimentation. The transcript of each video is processed to extract the context information present in it. This context information is further used to crawl the web to search for other relevant news content using this ontology-based approach. A corpus of around 125–150 documents consisting of a mix of videos and text files were obtained by crawling the web for each video. Out of the 100 videos, around 88 videos were tagged correctly with the relevant context and content information.

A record of temporal and geographical context for each video and text news is maintained in a file. The documents corresponding to these two domains were collected from the web over a period of two to three months. Concepts as evolved over time were tracked at a gap of three to five days. After applying pre-processing steps over the entire set of documents, LDA is run to obtain topic-document distribution. A total of 50 topics were considered. Highly probable words of each topic were independently mapped over the E-MOWL-based ontology. Each document was represented as a probability distribution over the concepts. Finally, a temporal pattern showing the conceptual evolution of various concepts over time is drawn. The trend is displayed in a two dimensional graph with probability of various concepts on the y axis and time on the x axis as shown in figure 8.11(a).

In the graph, the probabilities of concepts corresponding to four sub-events – namely, *Campaign, Polling, Results* and *Government formation* – for an Indian election event have been considered. The authors considered date of occurrence of subevents while analyzing the correctness of event patterns over time. The time at which the probability of a subevent is high was matched with the actual date of occurrence of that subevent. The time of the occurrence of subevent as observed in the graph was found consistent with the actual dates of their occurrences. The correctness of event patterns over spatial location was measured by manually analyzing the actual locations involved in that event.

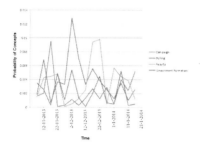

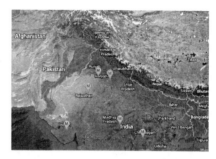

(a) Conceptual evolution over time

(b) Geographical/spatial trend analysis

FIGURE 8.11: Results from Delhi Legislative Assembly Election 2013

Geographical trend analysis using the ontology is done by obtaining the geographical event patterns in the video. Initially NLP was used to extract the geographic entities from the *Document Set*. Geo-ontology was used to establish the relation between these geographic entities and to provide semantics of the geographical schema over Google Maps, used as the low-level GIS. An example geographic trend corresponding to a *Movie Release* event is displayed over a Google Map in fig. 8.11(b). The Google map shows the various places which are related to the *Movie Release* event in some way.

8.4.3 News Aggregation from Sports and Political Videos

The previous sections are primarily based on text analysis using natural-language processing (NLP) and machine-learning techniques. Now we move to the realm of video data analysis using multimedia ontology. In [83], the authors used a MOWL based representation of knowledge pertaining to the domain of sports and political news. The probabilistic inferencing scheme of MOWL with Bayesian networks is used to infer the probability of occurrence of domain concepts given the observation of associated features and other related concepts in news videos. Semantic annotation based on content analysis of the videos is done using this inferencing framework. It is then stored in appropriate structures so that it can be used effectively for providing conceptual access to the video collection. The efficiency of this video annotation scheme is illustrated with the help of a browsing and querying interface that makes use of Synchronized Multimedia Integration Language (SMIL) representation.

8.4.3.1   Semantic Annotation of Video

The annotation framework utilizes the spatio-temporal characteristic of the audiovisual data of the video. A comprehensive annotation scheme takes care of video analysis at the multiple levels of granularity at which the entities exist and interact in spatial and temporal context. The scheme can associate annotations with a spatial region within a frame, a complete frame, a spatio-temporal region spanning a set of frames, a set of frames within a shot, a complete shot, a set of shots, or the complete video. Events within the video can also be specified at multiple levels of granularity: a spatial event where two entities share a specified spatial relation; a temporal event when two temporal entities exhibit a specified temporal relation; spatio-temporal events corresponding to the change of state, such as a change of spatial relation or temporal relation with another entity; and so on.

```
<Video source='football-034.mpg'>
    <VideoShot ID='1', t1='10.0', t2='190'>
        <ConceptEntity ID='12' t1='10', t2='28', locationType='SpatioTemporal', prob='0.78'>
            <BoundaryPolygon t1='12'>
                178 288   166 280 ....
            </BoundaryPolygon>
            <BoundaryPolygon t1='13'>
                ....
            </BoundaryPolygon>

        </ConceptEntity>
        <ConceptEntity ID='16' t1='45', t2='190', locationType='Temporal', prob='0.925'/>
        <ConceptEntity ID='17' t1='110', locationType='Spatial', prob='0.95'>
            <BoundaryPolygon t1='110'>
                178 288   166 280 ....
            </BoundaryPolygon>
        </ConceptEntity>
    </VideoShot>
    <VideoShot ID='2', t1='191.0', t2='212.0'>
        ....
    </VideoShot>
</Video>


<ConceptEntityStructure ID='12', name='GoalScore', definedInOntology='football.mowl'>
    <instance videoSource='football-034.mpg', shotID='1', locationType='SpatioTemporal', prob='0.78'>
        <BoundaryPolygon t1='12'>
            178 288   166 280 ....
        </BoundaryPolygon>
        <BoundaryPolygon t1='13'>
            ....
        </BoundaryPolygon>

    </instance>
    ............
</ConceptEntityStructure>
```

FIGURE 8.12: Structure of the semantic annotation for videos.

The semantic annotation is a triplet <*concept, location, probability*>, where *probability* specifies the likelihood of *concept* found at *location* (identifying a spatio-temporal region) in the video. As a default, the probability of the concept being relevant to the location is taken as 1 (one). The MOWL ontology-derived Bayesian network can now be considered to model the dependencies between concept hypotheses. It makes use of the evidential observations at the hypothesized locations and infers probabilities for the relevance of the concept/s. As a result of Bayesian network inferencing, the nodes that accumulate a marginal probability greater than 0.5 depict annotations that are relevant to the shot or scene.

The annotation structure proposed in this work, supports two ways of viewing a video: (a) linear viewing mode, which follows the inherent temporal order, with semantic annotations displayed and updated with the video progressing; and (b) abstract viewing mode, where abstractions like summary, scenes, events and object interactions can be viewed on demand. To support such an interactive access to video content, the video needs to be annotated at the syntactic and semantic levels – syntactic structure to represent video information like features or objects and semantic structure for story and information about its semantic entities like objects and events. Conceptual annotations are embedded in a basic annotation structure, which has a syntactic structure with semantics embedded within the appropriate level (shot, frame, or the complete video). Figure 8.12 shows a sample XML annotation for a video. The video is divided into shots that have several `ConceptEntities`, each of which is then linked to a concept in the ontology along with the likelihood of its occurrence at the specified location.

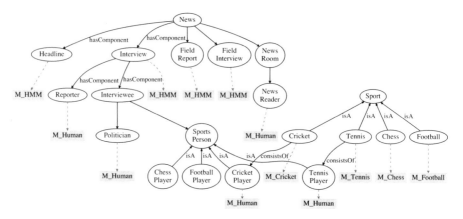

FIGURE 8.13: An ontology for the sports and news domain.

A domain ontology is required to identify the domain concepts for the annotation. Figure 8.13 shows an ontology for the domain of sports and news videos. A news video is considered to be classified into five segments:

- Headline sequence: The text of the news headlines is shown on the screen, often accompanied with music.

- News Room sequence: This shows the news reader reading news in the studio.

- Interview: This shows the interview taken indoors within the studio.

- Field Report: This shows the on-the-field news coverage.

- Field Interview: This shows interviews taken at outdoor public places.

A number of nodes in this ontology are from the sports domain, shown with their associations to the news-related concepts along with some media-based leaf nodes shown as dotted rectangles. These observable nodes provide evidence in the observation model towards recognizing a concept and identifying conceptual annotations for the video.

8.4.3.2   Sports and Political News Aggregation

The news aggregation application developed as part of this work illustrates the advantage of using a MOWL-encoded multimedia ontology to model the domain knowledge and combine it with semantic markup of multimedia data to offer an ontology-based querying and browsing interface. The application uses a SMIL-based representation of videos and a SMIL presentation engine to view and interact with these files. Two MOWL-encoded ontologies are used: ont-A, which encodes general scene characterization knowledge; and ont-B, a sports and news ontology shown in figure 8.13. The application uses trained Hidden Maorkov Model (HMM)-based classifiers that segment news videos and label each segment with one of the five news-scene categories. This video segment with a news-scene class label is evaluated to instantiate the corresponding scene type node in the Bayesian network for ont-B. Ontology ont-B also shows other scene classes like tennis, cricket, football, and so forth, which also occur in ont-A. The scene type nodes that are common in the two ontologies are instantiated in ont-B with the same inferred probabilities as obtained from Bayesian inferencing in ont-A.

The application achieved news aggregation from a video collection of over 215 video clips, processed to identify the blob-tracks and perceptual clusters using the DSCT and perceptual grouping algorithm respectively. Video clips were annotated using observation models derived from the two ontologies, ont-A and ont-B. The news aggregation engine makes use of the semantic annotations and offers SMIL-based presentation of videos. Users can navigate the collection by dynamically generating the hyperlinks depending on which concept is active, resolve queried concepts into related concepts if required, and retrieve video segments relevant to the query. Figure 8.14 shows two snapshots from the user interface, where the main query is `Football`. The user is shown related concepts for the selected video, from which she first selects `GoalScore` and is shown videos with goals being scored, and then selects related concept `Sports`, which causes aggregation of videos belonging to other sports like chess and cricket as well.

8.4.4   Discussions

Information aggregation from web resources becomes relevant in the context of the vast amount of heterogenous data that exists on the web and is localized with different web applications and portals. In section 8.4, we have shown how semantic interpretation of media contents can be used to effectively

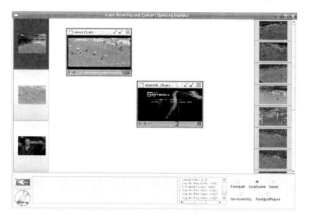

(a) Query Football with related concept GoalScore

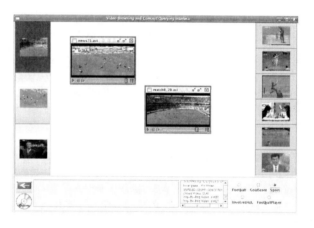

(b) Query Football with related concept Sports

FIGURE 8.14: Video retrieval results for different queries.

integrate and aggregate information, thus rendering better satisfaction of the user's information need. Multimedia ontology helps to establish the context of semantic information and also provides the much-needed associations between semantic concepts and media features. MOWL representation and its various extensions like event-based MOWL (E-MOWL) and Geo-ontology have been used to analyze spatio-temporal patterns present in news videos and event trends in text and video documents. Its probabilistic reasoning framework allows for recognition of event patterns based on full or partial evidence collected for subevents. Multimedia information aggregated in different media formats (i.e., text, video, images and maps) and presented with ontology-based context to the user makes for interesting audio-visual presentation.

8.5   Conclusions

In this chapter, we have shown a few applications that exploit the semantics of multimedia contents to identify multimedia events from different types of resources and to relate them to physical space and time. In this context, we have presented an extension to MOWL that can encode the spatial and temporal properties of an event. These spatial and temporal properties are with respect to the real-world coordinates and not the relative spatio-temporal organization of an event components, which were described in chapter 5. Many of the applications deal with heterogeneous media collections distributed over the net. We present a collaborative agent-based architecture that can be used to interpret the contents of such distributed collections with a common multimedia ontology in chapter 9. Heritage preservation involves preservation of digitized tangible heritage resources like monuments, handicrafts and sculptures, and preservation of intangible resources like language, art and culture, music and dance. It is a specialized, complex field and requires a knowledge-intensive approach. Application of ontology in heritage preservation allows for contextual and knowledge-based access to the digital artifacts, as well as their interlinking to provide cross-modal retrieval. Some of the applications in this domain are discussed in chapter 10.

Chapter 9

Distributed Multimedia Applications

9.1 Introduction

The use of multimedia data has become quite widespread on the Internet in recent times. It is estimated that multimedia data accounts for about 60–70 percent of total traffic on Internet and mobile networks. The applications include infotainment, remote surveillance, teleconferences, social networks, and education, to name a few. With the availability of commodity data capture and networking devices, the opportunity of authoring and sharing multimedia data has been extended to the masses, resulting in a huge explosion of multimedia on the web. It is estimated that 7 petabytes[1] of photos are added to Facebook alone every month and that 2.5 million hours of news video was added to Youtube during 2012 [157]. Much of such data is distributed over myriads of network nodes and on cloud-computing platforms.

The large volume of multimedia data from distributed data sources often needs to be analyzed and meaningfully integrated for its effective use by human beings. Examples of such applications are news aggregation from multiple sources [37, 10], rapid diagnosis using wearable device [58], personal photobooks [160], and digital heritage applications [146, 152]. Many of the applications need to process large volumes of multimedia data from several distributed resources in near real time.

In the following sections, we review the approaches to solve the challenges posed by web-scale multimedia data processing. Various issues of *big multimedia data* have caught the attention of researchers very recently. In this chapter, we concentrate on semantic integration of distributed and heterogeneous multimedia data. We briefly review the impact of multimedia data processing on the architectures of multimedia data cloud and cloud-based multimedia applications. In the context of knowledge-based multimedia data processing, we review the integration of knowledge distributed over the web. Further, we analyze how ontologies expressed with Multimedia Web Ontology Language can be used for semantic integration of distributed information resources. We present a possible agent-based architecture that can be used in this context and illustrate that with an application example.

---

[1]1 petabyte = 10^{15} bytes.

9.2   Challenges for Web-Scale Multimedia Data Access

Web-scale access to multimedia data poses quite a few unique challenges. The first and foremost of the challenges is, of course, the scale. Processing the entire data on the web for creating an inverted index is an intractable problem. Google claims to have indexed 10 billion images as of 2010, but it can well be imagined that many more have been left out.

Besides the large volume, another problem with multimedia data processing is the heterogeneity of media data, media applications, and the resources hosting them. The different media forms, such as still images, animation graphics, videos, and music, require different content processing techniques. The media properties, such as resolution, color depth, frame rate, sampling frequency, and signal-to-noise ratio, may vary from instance to instance and may convey information at different levels of granularity. The media data can be characterized through different feature descriptors, which is often guided by the specific needs of the hosting portals. The heterogeneity of metadata organization and local index structures also causes significant hurdles for web-scale media data analysis.

Distribution of multimedia data over the Internet also poses a significant challenge. Media data files are generally bulky. Once media files are discovered through crawling, transferring them to a processing node for their analysis for example, feature extraction, poses a large overhead burden on the network. Copying media data across a network may also conflict with data privacy requirements. While a media archive may be willing to expose a teaser to attract potential users, it might not like to allow unrestricted access to its complete data.

9.3   Architectures for Distributed Multimedia Data Processing

Multimedia systems in the 1990s were typically designed to process a few thousand media instances, typically images, stored on a desktop machine. Federated systems, where search was conducted independently on the different nodes and the results were subsequently merged, were used for distributed media retrieval [39, 16]. In these approaches, the semantics of media data were locally ascertained at the participating nodes without a global consensus, which posed a challenge in merging the search results from the different nodes. Grid computing infrastructure has been used in [31, 172] to solve the computing bottleneck for media content analysis, though at the cost of increased communication overheads in transferring large media files across the network. Specially designed hardware like the graphic processing unit (GPU), provide fast access to memory and parallel processing of large blocks of data to speed up multimedia data processing. Common media analysis tasks, such as feature extraction, K-means clustering, and KNN search, for large volumes of media data can benefit from the massive parallelism in the distributed processing system architecture, employing Map-Reduce software architecture [46] and Hadoop Distributed Filing System (HDFS). Examples for such algorithms can been found in [238, 232, 6]. A method for concept modeling with a large volume of media data over a parallel processing platform has been proposed in [225].

In the current decade, there has been a major paradigm shift in the way multimedia data is acquired, stored, processed, and consumed. The shift is primarily driven by the availability of affordable data-capture devices like digital cameras and smart-phones, which have empowered the public to capture images, music, and video. Numerous media-sharing and social networking sites, like Facebook, Youtube, Instagram, and Flickr, support sharing of such media artifacts over the web. Economic processing and storage of the large volume of media data have been supported by cloud computing platforms. This has motivated several organizations, such as National Geographic and the Smithsonian Institute, to create large multimedia data archives in the cloud.

While the cloud provides the flexibility of large processing power, and on demand storage, and economy of scale, multimedia services in the cloud need some special attention. This is especially important in view of limited processing power and connectivity of the mobile handsets that are likely to become major consumers of such services. The quality of service (QoS) heterogeneity of the access devices needs on-demand adaptation of multimedia data. Zhu et al. [237] argue a case for multimedia-aware design of clouds and cloud-aware design of multimedia service architecture. In a multimedia-aware cloud architecture, multimedia data are placed on the edge of the cloud, closer to the end users (possibly with redundancy), so that users can have fast and efficient interaction with the data. Each of the edge nodes can indeed be a cluster of high-performance processors with adequate storage to enable complex processing of large volumes of data.

While the data are distributed on the various clusters in the cloud, there needs to be some mechanism to locate them. A traditional mechanism is maintenance of a central index, as shown in figure 9.1(a). The major benefit of this architecture is that a user needs to consult a single reference point to access any data. However, there are many pitfalls of this scheme. Creation of such an index requires a significant volume of media data to be copied across the network, which can prove to be a bottleneck in arbitrarily scaling up such systems. Moreover, exporting data for indexing may contradict the data privacy requirements or commercial interests of some of the repositories. As a result, the data buried in such repositories will never be indexed. On the other hand, there may be hanging pointers when obsolete data in the distributed nodes are deleted. Other major issues in central indexing are that the indexing node needs to deal with data in heterogeneous formats and media forms and that semantic indexing may not benefit from the specific knowledge about the collection.

As an alternative to the central indexing scheme, Zhu et al [237] propose a peer-to-peer architecture, where a set of processing nodes are colocated with media data clusters in a cloud-computing environment. A data cluster together with its processing node becomes an autonomous unit that can process a user's request locally. The processing at a node can be adapted to the specific requirements for the nature of media data and its organization in the cluster. The independent units collaborate at the peer level to facilitate data access from the distributed collections, as shown in figure 9.1(b). We shall build on this architecture in the following sections.

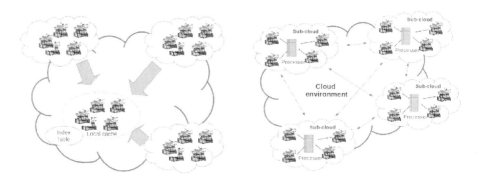

(a) Central indexing architecture   (b) Peer-to-peer architecture

FIGURE 9.1: Alternate architectures for multimedia data access.

9.4   Peer-to-Peer Architecture

Early peer-to-peer (P2P) systems like Gnutella, BitTorrent and Kazaa have been extensively used for file, music, and video sharing. Every participating client in these systems acts also as an information resource and as a server. It not only proves to be a scalable system but also economically viable, since many distributed commodity machines are used as the computing platform. While data access in these systems is generally through annotations, (e.g., artist or the album), a content-based music retrieval system based on the *query-by-example* paradigm over a P2P architecture has been proposed in [228]. The P2P systems alleviate communication overhead in a distributed multimedia system through local data processing.

There have been several attempts to improve the performance of the basic P2P architecture. Vlachou et al. [197] have proposed a semi-P2P architecture for a content-based similarity search where a hierarchical indexing scheme has been used to route feature-based queries to promising information sources through a set of broker nodes. Crespo and Garcia-Molina [38] have proposed a multilevel P2P network architecture, where the nodes with semantically similar contents, as determined by metadata, are interconnected at the lowest levels to reduce communication overhead in the system.

9.5 Multiagent Architecture for Distributed Multimedia Systems

A multiagent system is a convenient tool for modeling peer-to-peer architecture. It comprises several *autonomous* pieces of software (agents) distributed over a set of computing nodes. Each of them acts independently and attempts to fulfill its own goal. The agents interact with other agents in a system at the peer-level. The interactions lead to collaboration that helps in realizing the system functionality, as well as ,to competition for the agents to maximize their individual gains. Behaviorally, a multiagent system emulates a human society. Multiagent systems have been used to implement large and distributed information systems with heterogeneous and independent information sources [173].

A typical architecture of a multiagent distributed multimedia information system is shown in figure 9.2(a). The system provides access to multimedia contents from a number of large heterogeneous media collections. Each of the resources is represented by an autonomous *resource agent* that controls access to its contents. A resource agent possesses knowledge about the collection; it autonomously indexes and facilitate access to a media collection. A class of agents, called the *user agents*, creates a user model and implements user interfaces over a variety of interaction devices. Another class of agents, the *broker agents*, facilitates interaction between the user agents with the resource agents. The broker agent interprets the user's request and breaks it down to smaller information-gathering tasks and subcontracts the tasks to different resource agents, depending on their capabilities. The broker agent integrates the partial results from the resource agents before reverting to the user. In general, there can be some redundancy in web resources. The redundancy may be in terms of data, (i.e. same or similar media instances are available at multiple sites), or in terms of the access mechanism, that is more than one gateway implementing different index structures or access methods are available for the same set of data. The broker agent negotiates with the resource agents to select an appropriate subset of resource agents in the context of a user's request.

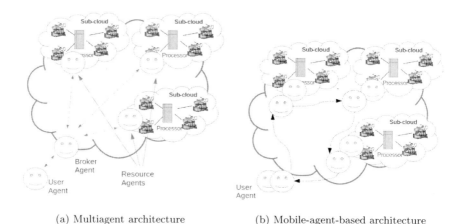

(a) Multiagent architecture   (b) Mobile-agent-based architecture

FIGURE 9.2: Agent based architectures for distributed multimedia information system.

Mobile agents are a class of agents that can move from one network node to another. They carry their state as they move from one node to another. A typical architecture for a mobile-agent-based multimedia information system is shown in figure 9.2(b). A mobile agent launched from the user node traverses some resource nodes. During the traversal, it interacts with the resource agents to gather information and incrementally aggregates the gathered information. Finally, it returns to the user with accumulated information. In a minor variation, a user agent launches multiple mobile agents, each of which traverses a different set of resource nodes. The user agent aggregates the information collected by the different mobile agents. This approach results in parallel information gathering and is resilient against failure of a mobile agent. Path planning (i.e. the information resources to visit and their order), is an important consideration in designing mobile-agent-based information systems. Termination criterion — when the mobile agent should discontinue its information gathering task and return to user node — is another issue.

9.5.1   Advantages of Agent-Based Architecture

Klusch et al [114] argue several merits for an agent-based architecture for distributed information systems. The main advantages are as follows:

- *Scalability.* Inherent parallelism in agent-based systems makes them scalable. Peer-to-peer architecture (as opposed to a central indexing structure) makes them scale better over a larger number of nodes.

- *Autonomy of information agents.* The different agent types in the system execute autonomously, with different knowledge resources and independent of each other. Different processing strategies can be applied at the different information sources in compliance with its data characteristics, organization, data security, and any other policies.

- *Independence and proactiveness.* Each agent may act independently with no or little human interaction in overcoming any obstacles, such as resource constraints, during their operations. Each agent proactively determines its goals driven by user preferences and information gathering needs, and works towards fulfilling them.

- *Dynamic selection of resources for data gathering.* The contents of the information resources on the Internet are generally dynamic in nature. An agent-based system enables dynamic selection of information resources in the context of specific information requirements and the current state of the resources.

- *Data Security and Trust.* The agents may expose only a part of the information that they deal with to the other agents as required in a specific task context, enhancing data security and alleviating trust issues. However, unreliable, insecure, or hostile information nodes can be a threat to the mobile-agent-based architecture. The architecture proposed in figure 9.2(a) performs better in this aspect.

9.5.2 Agent Coordination and Ontology

In a multiagent system, the agents need to coordinate with each other to realize the system capabilities. Thus, *social ability* — the capability to effectively communicate and interact with other — is needed. FIPA standards[2] deal with the syntactic layer of the communication. Further, all agents in an agent team need to have a uniform semantic interpretation of the communicated message tokens. This can be achieved with a shared knowledge structure, or an *ontology* that describes the domain of discourse. Such an ontology provides a shared conceptualization of the domain, or a shared virtual world, where the agents can ground their beliefs and actions [97]. In the context of collaborative multimedia data analysis, a common understanding of the nature of the media features, operations that can be performed on them, and their relations with the concepts in the domain of discourse by all participating agents is essential.

---

[2]www.fipa.org.

9.5.3   Ontology for Multimedia Data Integration

The heterogeneity of data sources and media forms poses a major challenge to semantic integration of multimedia data. Each of the media forms needs to be processed differently for information extraction. In general, the information extracted from different media forms from a multimedia data-stream are complementary in nature. For example, while it may be possible to identify the genre of the music from the audio recording of a music performance, a still photograph may identify the musician. Similarly, the analysis of speech and the visual contents of a news videoneeds different approaches and may yield different types of information [68, 109]. Such complementary pieces of information need to be semantically integrated for identifying the high-level concepts in multimedia data [151]. A generic multilevel architecture for semantic interpretation of multimodal data is shown in figure 9.3. In this architecture, data in different media forms are processed with media-specific analysis tools to discover media features (at different levels of abstraction), which are subsequently interpreted with an ontology and then integrated. The ontology that is required for such multimedia data integration has many facets. It needs to encode the domain and application concepts, the spatial and temporal milieu of the events that it deals with, the objects in the multimedia instances, and the semantics of the media descriptors in the different media forms [137]. It also needs to encode the complex structural, spatio-temporal and other relations between these entities in the context of the application.

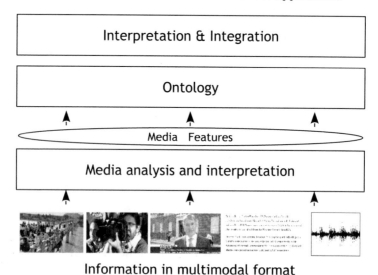

FIGURE 9.3: Framework of semantic fusion of multimedia data.

In this context, it may be recalled that there is no common way to interpret or to represent multimedia information contents. While MPEG-7 provides a standard for multimedia data description, an instance of multimedia data can be described in many different ways, each complying with the MPEG-7 standard. Thus, each of the independent multimedia data sources may use different encoding schemes to represent the data in its collection. Such encoding is generally guided by the local view of the data set that may or may not be governed by a formal domain description in the form of an ontology. As an example, social media generally depend on *folksonomy* which is a collaborati/ve and generally uncontrolled approach to categorize contents through social tagging.

9.6   Ontology Alignment

Integration of data from multiple information sources requires their interpretation from a common viewpoint. The viewpoint should further be aligned with the application and the domain. For example, the CIDOC ontology [49] has been created to interpret the metadata associated with multimedia artifacts available across a network of museums in Europe. The LSCOM vocabulary [138] attempts to encode all possible concepts that can be discovered in news videos. Development of such large ontologies requires an extensive ontology engineering exercise, involving selection of the *right* concepts and relations based on discussions and negotiations across all stakeholders. Besides the huge effort and coordination involved in developing such massive monolithic ontologies, enforcing an open set of agents that can potentially participate in a collaborative problem-solving effort to commit to such ontology is a major challenge.

As a example, consider the websites that host a large volume of crowd-sourced images and videos. The contents in these sites are complemented by a variety of categorization and annotations, such as captions, descriptions, tags, geo-tags, groups, and galleries. The different sites have been created with different goals and promote different types of annotations. For example, while Flickr aims at showcasing quality photographs to a global audience, the goal of Kodak Gallery[3] has been to share photographs of personal events with family and friends. As a result, there has been a significant difference in interpretation of images by the users of the two websites, which is reflected in the organization of these sites and the annotations styles of the users [143]. Thus, it is difficult to conceive a common ontology for the two disparate sites. Even for the same collection (like Flickr), researchers have proposed different ways to analyze the contents and the metadata to provide semantic access to the its contents [142, 64, 224, 212]. Each of these approaches creates a unique

---

[3]Merged with Shutterfly since July 2012.

ontological description of the collection without reference to any specific domain.

Thus, a pragmatic alternative to create a large monolithic ontology is to allow the information sources to use their own ontologies and to dynamically align them with a domain or an application ontology in the context of use. The problem of ontology alignment involves discovering the relations between the entities across the ontologies in an application context. Once the connections have been found, the different ontology fragments can be used together to solve the problem of semantic access to multiple media sources. Since the ontologies are independently developed, they may use different vocabulary as well as different relations between their concepts. As a result, the problem of ontology alignment, or dynamic integration of ontologies, is a nontrivial one.

9.6.1   Definitions

The problem of ontology alignment can be loosely defined as finding correspondence between the entities across two or more ontologies. The entities can be of different natures — for example, the concepts, the instances, the relations and even a subgraph. In the strictest sense, "correspondence" may mean strict equality, but it may also be extended to include different types of relations, such as subsumption, concatenation (e.g., *name* = *first-name* + *surname*), equality with value restrictions (e.g., an *auto-rickshaw* is a *taxi* with three wheels), quantitative and qualitative descriptions (e.g., slow car meaning speed < 40 kmph), and so on. Ehrig [51] defines general ontology alignment as a partial function:

$$genalign : E \times O \times O \rightharpoonup E \times M$$

where E represents vocabulary of all terms $e \in E$, O represents the set of all possible ontologies, and M represents the set of all possible alignment relations. Often, the results of ontology alignment are not crisp, but are associated with a confidence level. For example, the confidence level in mapping a height of 178 cm in one ontology to *tall* may be marked with a confidence value of 0.9.

The concept "similarity" plays a keyrole in ontology alignment. Loosely speaking, similarity between parts of two ontologies is a measure of the degree of correspondence between its constituent elements. Formally, it can be defined as a function [51]:

$$sim : \beta(E) \times \beta(E) \times O \times O \rightarrow [0, 1]$$

where $\beta(E)$ represents a set of entities in the ontologies. The similarity function results in a real number between 0 (least similar) and 1 (most similar).

9.6.2   Similarity Measures

The similarity of ontological entities is established in multiple layers [51]:

1. **Data Layer:** At the lowest level, the data values are compared. Examples include comparing the data labels (strings in a given name space) or the URLs that they may be pointing to. Complex data types may need more sophisticated computations, including conversions and approximations (e.g., $5'10''$ vs. 178 cm).

2. **Ontology Layer:** At the next level, the semantic relations between the entities are explored. It utilizes the ontology graph, the semantics of the relations, and the property restrictions.

3. **Context Layer:** At the highest level, the similarity is determined based on factors external to the ontology, such as an application context. Here, the notion of similarity comes from common usage in similar contexts. For example, if users with similar profiles commonly like two movies, they are considered to be similar.

It may be noted that the three layers are not independent but influence each other. The similarity found in the data layers reinforces similarity computation in ontology and context layers and vice versa. As a result, the similarity computation generally iterates across these layers. Another important aspect of similarity computation is *feature engineering*, which suggests the properties of the entities that should be considered for establishing similarity. For example, two *concepts* in two ontologies may be considered identical if they have equivalent labels as identified by a linguistic tool, like a dictionary or WordNet; for example, an Anglo-French dictionary may be used to equate the concepts "car" and "voiture" in two different ontologies. As another example, two *instances* in different ontologies may be considered equivalent if they point to a common resource, that is if they share a common URL. The different relations in WordNet can be used to establish other types of alignment relations, such as subsumption and whole–part relations.

9.6.3   Similarity Measure with Media Features

The entities in an ontology are generally expressed with linguistic tools and may look dissimilar because of the use of different language or different terminologies. An interesting approach to alleviate this problem is to establish relations between entities in different ontologies by comparing their media properties. The basic premise behind the proposition is that the same concepts should have similar media properties. For example, all cars, irrespective of what they are called, share a common body shape. As another example, music of the same genre generally has similar tempo and pitch. Perceptual modeling of a domain using MOWL presents such an opportunity. It is possible to derive an observation model of a concept from an ontology encoded in MOWL. The observation model characterizes a concept with its expected media properties, which hold good for different instances of the concept. Thus, two concepts can be said to be equivalent if the observation models for two concepts are similar [127]. Further if the observation model of a concept B is a specialization of that of concept A, the concept B can be said to be subsumed in concept A. Note that the observation model of a concept incorporates media properties of related concepts in a domain. Thus, similarity of concepts is established not only at the data level, but also at the ontological level using this approach.

For example, consider the observation model for a class `Monument` derived from ontology O_1 that incorporates a few prominent architectural components like the domes, minarets, and facades as its expected media manifestations as shown in figure 9.4(a). On the other hand, the observation model for the `Tajmahal`, which is an individual in ontology O_2, incorporates a specific spatial arrangement of similar components, besides its color and texture properties, and an example image as shown in figure 9.4(b). The observation model of the `Tajmahal` is a specialization to that of `Monument`. Thus, it is possible to conclude that the individual O_2:`Tajmahal` is a member of the class O_1:`Monument`.

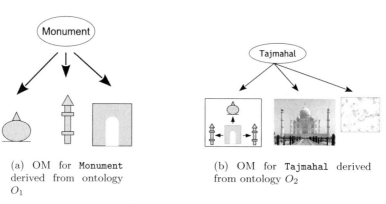

(a) OM for `Monument` derived from ontology O_1

(b) OM for `Tajmahal` derived from ontology O_2

FIGURE 9.4: Ontology alignment using observation models.

9.7   Knowledge-based Access to Massively Distributed Multimedia

In this section, we expand on the agent-based architecture for multimedia information systems, introduced in figure 9.2(a). The information system comprises a number of independent information sources, each hosting millions or billions of data instances. Each of the resources uses a local ontology to provide semantic access to its own collection. The semantics provided by the different resources may be quite different from each other and from that required by a user in a specific application context.

The contents from the information sources need a common interpretation despite heterogeneity for information integration to be possible. For example, integrating information on cultural events may require uniform categorization of event types, recognition of the specific event instance, and association of the event with other entities, such as artists and sociocultural implications, which may not be possible with the local ontologies. Thus, such information gathering would require intelligent task decomposition and allocation, enabled through ontology alignment and knowledge-based interactions between the participating agents. We show an approach to achieve such agent coordination using ontologies expressed with Multimedia Web Ontology Language (MOWL).

9.7.1 Observation Model, Minimalist Plan, and Observation Plan

MOWL enables concept recognition in media documents by creating an observation model (OM) for a concept. An OM comprises a set of media patterns that are expected when a concept occurs in a media instance. Observation of any of these media patterns in a media instance provides evidence that a concept appears in the media instance. Concept recognition relies on cumulative evidence as an observation of a number of such media patterns. The reasoning with media properties of related concepts in the MOWL reasoning framework helps incorporate conceptual cues in the observation models. For example, an OM for a dance recital might include the characteristics of the music that is likely to accompany the dance form. An OM is organized as a Bayesian network, where the root node represents the concept being explored, the intermediate nodes represent some intermediate concepts or media objects, and leaf nodes represent the media patterns to the observed. The leaf nodes are also called the observation nodes. An OM created from a nontrivial MOWL ontology generally comprises a large number of media patterns, only some of which need to be observed to attain an acceptable belief in the existence of the concept being explored. This redundancy in an OM provides an opportunity for the resource agents to choose different subsets of media patterns from the set prescribed in the OM for concept recognition. This implies that the different resource agents can apply different methods for concept recognition under a common ontological framework.

Given an observation model for a concept, a resource agent needs to create an "observation plan" for concept recognition for the documents in its collection. An observation plan is defined as a subgraph of the observation model, that includes a subset of observation nodes, that when observed, creates sufficient confidence in the discovered concept and at the same time, should be computationally and economically efficient. Creation of such a plan requires interaction of two independent forms of knowledge — namely, the domain knowledge that relates media patterns with the concepts and the collection knowledge that describes the idiosyncrasies of the collection.

A distributed planning algorithm for creating such plans with participation of the broker and the resource agents is proposed in [67]. The planning is performed in two stages. In the first stage, the broker agent reasons with an observation model to create a "minimalist plan" that is a subgraph of the OM with the minimum possible number of observation nodes with which it is possible to achieve a given confidence target for the concept, that is a given posterior probability of the root node. The minimalist plan does not consider the feasibility or the cost involved in making the observations at the different information sources. In the second stage, the minimalist plan is adapted to observation plans for the information resources to meet some specified cost constraints. At this stage, some observation nodes with an associated high cost of observation may be replaced by others. Since the cost of observing a media pattern is specific to an information resource, the observation plans are resource specific and their creation is guided by the collection knowledge.

9.7.2   Formulating Minimalist Plan

A "minimalist" observation plan comprises a nonredundant subgraph of the OM that can fulfill a confidence target. This means that when all media patterns that are specified in the observation nodes of a minimalist retrieval plan are discovered in a media instance, the root node will attain a posterior probability greater than or equal to a specified value. But if any of the observations are skipped, the posterior probability cannot be attained. There can be several solutions to the problem in a minimalist plan.

The change in posterior probability of the root node in a Bayesian network on instantiation of a leaf node depends on the network topology, the conditional probabilities, and the states of instantiation of the other leaf nodes. Thus, the number of observation nodes required in a minimalist plan depends on the nodes chosen. An optimal minimalist plan is the one with the minimum number of observation nodes. Finding a general solution to the problem requires examination of all possible combinations of the observation nodes and is of the complexity $O(2^N)$. Algorithm 7 depicts an alternative way for computing a minimalist plan. It is essentially a greedy algorithm that selects the observation nodes successively in order of their decreasing influence on the root node, computed dynamically in the context of other observation nodes already selected. The algorithm stops when the root node can attain a posterior probability greater than the specified value as a result of observations on selected observation nodes. If the desired posterior probability cannot be attained even after inclusion of all observation nodes, the OM is considered inadequate for the information-gathering problem. At any stage of computation, the most influential observation node is selected by following the most promising path starting from the root node. The algorithm is guaranteed to find a solution to the problem, if it exists. In the worst case, the solution can be suboptimal.

Algorithm 7 Algorithm for finding a minimalist plan

Inputs: OM with no paths marked; confidence target (t)
1: Initialize Minimalist Plan (M)   ▷ $M = \emptyset$
2: **while** (there are some more observation nodes) **do**
3:     Find the most influential observation node n in the unmarked graph
4:     Mark the path π from the Root node to n; $M = M \cup \pi$
5:     Instantiate n; find posterior probability p at the root node
6:     **if** ($p > t$) **then**
7:         Return M ▷ Confidence target met; M is the Minimalist Plan
8:     **end if**
9: **end while**
10: return $FAILURE$ ▷ Solution to minimalist plan does not exist

---

9.7.3 Formulating an Observation Plan

The minimalist plans so created do not consider the capability of the resource agents to make the selected observations. Some of the observations may be too costly in terms of computing requirements or impossible for a resource agent. For example, it may be quite expensive to perform face recognition in some surveillance video collections and impossible to observe some audio patterns in an image repository. We use the term "cost" in a generic sense. For example, it may represent computational complexity when computational infrastructure is at a premium, or it may be a monetary cost when there are financial implications in scheduling the media pattern-recognition tasks.

Thus, the minimalist plan needs to be adapted for an information resource for feasibility and to conform to some cost constraint. The adapted plan is called an observation plan for a resource. Since the cost for observation of a media pattern generally depends on the resource organization and hosted media forms, the observation plan needs to be resource specific. The broker agent collaborates with the resource agents to adapt the minimalist plan to observation plans for the respective resources.

Algorithm 8 is used to adapt a minimalist plan to an observation plan for a resource agent. The resource agent estimates the costs for the individual observations in the minimalist strategy. If the total cost exceeds a specified cost constraint, the costliest observation node is deleted from the plan. Since a minimalist strategy does not have redundancy, the confidence target can no longer be reached as a result of such deletion. In order to regain the confidence target, some observation nodes are to be added to the plan. The algorithm treats the observation nodes in an OM that were not selected in a minimalist plan to be in reserve for the purpose. One or more nodes from the reserve are added to the plan till confidence target is regained. The reserve nodes are selected in order of their decreasing influence on the root node as in section 9.7.2. The total cost is checked again and if it still exceeds the constraint, the cycle repeats. The algorithm stops when a solution within the specified lower bound of confidence target and upper bound of computation cost is found. The solution is the *observation plan* for the resource. If such a solution cannot be found for a resource, the corresponding resource agent cannot participate in the concept recognition task. Note that though the number of observation nodes in an observation plan may be more than that in a minimalist plan, the total cost of observations have decreased since the combined cost of observation for the added nodes can be less than that of the deleted nodes.

Algorithm 8 Algorithm for adapting minimalist plan to observation plan for a resource agent

---

Inputs: Minimalist Plan (M); Reserve nodes (R); Cost target (C); Confidence target (t)

1: Copy M to trial Observation Plan P
2:                  ▷ Trial Observation plan is same as the minimalist plan to begin with
3: **while** () **do**
4:      Compute total observation cost (c) of P
5:      **if** ($c < C$) **then**
6:          Return P ▷ Cost and Confidence targets met: P is Observation Plan
7:      **else**
8:          Remove costliest observation node from P
9:          **while** ($R \neq \emptyset$) **do**
10:                  ▷ Keep on adding nodes from reserve till confidence target met
11:             Find the most influential observation node $n \in R$
12:             Add n to retrieval plan; remove n from R
13:             Instantiate n; find posterior probability p at the root node
14:             **if** ($p > t$) **then**
15:                 Break   ▷ Check if cost constraint satisfied
16:             **end if**
17:          **end while**   ▷ No more observation nodes in reserve
18:          Return $FAILURE$ ▷ Solution to observation strategy does not exist
19:      **end if**
20: **end while**

The resource agents, in turn, may subcontract some complex media processing tasks (e.g., face recognition) to other agents, when such services are publicly available on the net. In general, several alternative media-processing resources may exist for a certain media processing task, each with a different quality and cost implications. A resource agent may explore such options and may make an optimal bid (or several alternative bids) during the negotiation process. At the end of this planning stage, contracts are established between the broker agent and a set of resource agents that can satisfy the requirement. Separate contracts may be established between each of the resource agents and a set of media processing agents.

9.7.4 An Illustrative Example

Consider an observation model for Bharatanatyam, an Indian classical dance form as depicted in figure 9.5. It comprises several observation nodes with media patterns in image, audio, video, and text forms. The observation nodes are marked with numbers 1 to 11 for reference in the following text. The conditional probabilities are not shown.

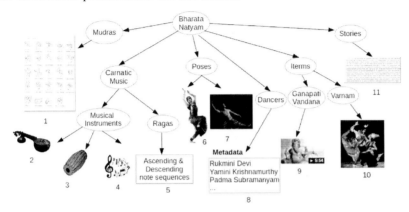

FIGURE 9.5: An observation model for Bharatanatyam.

The dance form can be recognized in a media instance by observing some of the media patterns in a media instance. However, observation of the different media patterns lead to different belief values in the concept. For example, while observation of a pose may lead to a high belief in the performance of the dance form, confirmation of a story that is common to many other classical dance forms will have least effect. The minimalist plan selects a set of observation nodes, which have the largest causal strengths for recognizing the concept. Assume that these nodes are 6, 3, 9, and 4 in order of their decreasing causal strengths. This selection depends only on causal strengths of the observation nodes and disregards feasibility or cost of observing the media patterns in an information resource. Figure 9.6(a) shows the minimalist plan.

Now, let us consider an information resource that contains a collection of manuscripts on the subject in digital form (document images). These manuscripts contain text and illustrations. It is not possible to discover an audio or a video evidence in these media instances. This implies an infinite (very large) cost for node number 9 in the minimalist plan. Thus, this node is deleted in the observation plan for the resource and some other node — say 8 (metadata) — is added to compensate and regain the confidence target. Similarly, node 4 (audio) is also replaced with, say, node number 11 (text). At this stage, the observation plan will contain the observation nodes 6, 3, 8, and 11. Assume that the total cost of computation for the observation plan is still higher than the specified cost constraint. Node number 6 is likely to be the node with highest cost, since it may be computationally expensive to determine a dance pose. Thus, node 6 is now deleted and, say, node number 1 (sketches) is included to regain the confidence target. The observation plan contains the observation nodes 3, 8, 11, and 1 at this stage. If this configuration of observation plan meets the cost constraints, it is accepted for the resource. Figure 9.6(b) shows the observation plan. This observation plan is specific to a document image collection; an observation plan for a video library could be quite different,

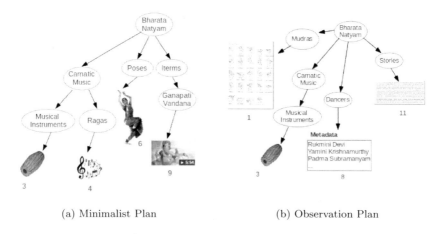

(a) Minimalist Plan   (b) Observation Plan

FIGURE 9.6: Planning for concept recognition.

The media features specified at observation nodes need common interpretation by the broker and the resource agents. The common understanding of the low-level media features can be achieved by deploying a consistent way to specify the media features, for example, by using MPEG-7 descriptors. However, MOWL permits observation nodes to be specified at different levels of abstraction. In particular, an observation model may contain high-level media objects, such as a dance pose. Such high-level constructs may have different interpretations with different resource agents. This can be resolved by the approach of ontology alignment as described in section 9.6.3.

9.8   Conclusion

In this chapter, we explained the use of multimedia ontology together with an agent-based architecture to create web-scale distributed multimedia information systems. The autonomy of agents in an agent-based system facilitates such arbitrary scale-up. An interesting feature of this architecture is the separation between the knowledge-based resources and their interactions. In particular, the large multimedia resources continue to use their local ontology and optimized access mechanisms. This is in contrast with some existing large distributed multimedia systems [49, 151], where every data instance need to be annotated with a common ontological framework. In the proposed architecture, knowledge-based interactions between the broker and the resource agents result in a customized observation strategy to be created for each participating information resource for every user information need in a specific domain context. Since all observation plans are derived from the same domain model, they represent a common viewpoint, albeit being distinct, facilitating information integration across the repositories. Another interesting aspect of this architecture is that the same information resources, without any modification, can participate in many distinct domain specific information systems, where each of such system is characterized by a unique domain ontology.

Chapter 10

Application of Multimedia Ontology in Heritage Preservation

10.1 Introduction

The digital medium offers an attractive means for preserving and presenting tangible, intangible, and natural heritage artifacts. It not only preserves a digital representation of heritage objects against the passage of time but also creates unique opportunities to present the objects in different ways, such as virtual walk-throughs and time-lapse techniques over the Internet. In recent times, the economics of computing and networking resources have created an opportunity for large-scale digitization of heritage artifacts for their broader dissemination over the Internet and for preservation. Many virtual galleries[1] have been put in cyberspace by well-known museums and cultural heritage groups to interface with a global audience. Currently, most of such presentations available on these portals are hand-crafted and static. With the increase of collection sizes on the web, correlating different heritage artifacts and accessing the desired ones in a specific use-case context creates a significant cognitive load for the user. Of late, there have been some research efforts to facilitate semantic access in large heritage collections. Several research groups [100, 81, 193, 155] have proposed use of ontology in the semantic interpretation of multimedia data in collaborative multimedia and digital library projects. The ontology and the meta-data schemes are tightly coupled in these approaches, which necessitates creation of a central metadata scheme for the entire collection and prevents integration of data from heritage collections developed in a decentralized manner.

The significance of a heritage artifact is implied in its contextual information. Thus, the scope of digital preservation extends to preservation of the knowledge that puts them in proper perspectives. Establishing such contexts enables relating different heritage artifacts to create interactive tools to explore a digital heritage collection. A major activity in any digital heritage project is to create a domain ontology [182, 91, 3]. But as digital heritage artifacts are mostly preserved in multimedia formats, they need text annotations

---

[1]For example, the Web Gallery of Art(www.wga.hu), Kalasampda Digital gallery (www.ignca.nic.in/dlrich.html), Louvre Museum(www.louvre.fr).

for semantic processing by traditional ontology representation schemes. Manually generating these annotations is a labor-intensive process and a major bottleneck in creating a digital heritage collection. In this context, we propose the use of the Multimedia Web Ontology Language to encode ontologies in the heritage domain, so that they can be easily enriched with multimedia data and and allow for semantic annotation and interlinking of the heritage artifacts.

In this chapter, we discuss applications in the domain of heritage knowledge management using a multimedia ontology. We illustrate the advantage of using MOWL and its probabilistic reasoning framework in preserving tangible and intangible cultural heritage. The domain selected for intangible heritage preservation is Indian classical dance, an ancient heritage. The MOWL ontology-based framework tries to capture the various interpretations of the scholarly knowledge of classical dance in actual performances by dancers and dance-gurus over time as well as in the contemporary world. This is made possible by use of a MOWL-based ontology to encode the inter-dependencies between various facets of dance and music and by building a knowledge base enriched with multimedia examples of different aspects of dance. With the help of this multimedia enriched ontology, the system is able to attach conceptual metadata to a larger collection of digital artifacts of the heritage domain and, provide a semantically integrated, holistic navigational access to them. The "Nrityakosha" application is basically an encyclopedic reference to multimedia resources of Indian classical dance, made possible with the help of an ontology.

To augment physical or virtual tours of heritage sites, contextual information should be readily available to the tourist. Audio guides and visual cues help in this goal, but their use is not sufficient to completely satisfy the information need. In this chapter, we have proposed a new paradigm for exploring heritage using a multimedia ontology. The proposed paradigm offers an *intellectual journey* of a heritage site, with readily available tools to traverse in space and time and access e-heritage artifacts in context and on demand. This intellectual exploration can also be extended to an abstract heritage theme, like a mythological story, which can have historical and contemporary significance, and thus utilize the infrastructure of a space-time traversal offered by this paradigm. We have illustrated this paradigm with a framework that offers ontology-guided navigation and a dynamic display of artifacts in context of a theme related to the World Heritage Site of Hampi in India. In another application, we have shown how such an ontology-based framework can easily be extended to offer cross-modal access to heritage resources preserved in different digital media formats.

10.2 Preserving the Intangible Heritage of Indian Classical Dance

Intangible heritage resources like music and dance forms are preserved by recording the performances of various artistes and exponents in multimedia format. The significance of such a preservation lies in the numerous stories and expressions that have evolved in the prevailing social, cultural, and geopolitical milieu and the styles of their depictions. An archive of heritage artifacts needs to make use of an ontology that *formally* captures the complex relations between such entities and their depictions to establish the context for the multimedia artifacts. Recent projects on digital heritage preservation [171, 49] use traditional ontology representation schemes to create a core ontology and to relate the metadata. Although they can reason with the domain concepts, they cannot reason with the multimedia representations of the heritage artifacts and their properties.

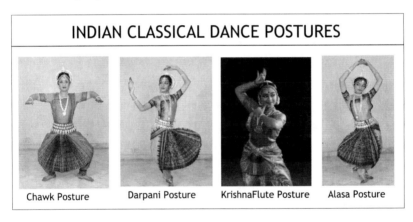

FIGURE 10.1: Some postures of Indian classical dance.

We have described an ontology-based digital heritage application called Nrityakosha in the domain of Indian classical dance (ICD) in our earlier work [129]. In this section, we elaborate on the need for MOWL to model the domain. ICD is based on a well-defined grammar for portraying mythical stories, for enacting the roles therein, and for choreographing dance sequences through a series of dance steps, gestures, and postures, some of which are shown in figure 10.1. The audiovisual depictions (e.g., the dance steps, gestures, and postures) in an instance of a dance performance are manifestations of the background concepts (e.g., mythical stories, episodes, roles, emotions). Concept recognition from the audiovisual recordings of dance performances requires the domain model to capture the relations between the concepts and their media manifestations. The media manifestations in a dance recital can be said to be *caused* by the background concepts that the artiste intends to portray, and thus provide evidence for the latter. Further, the individual dancers make their own experiments with the compositions and exercise some freedom, introducing a good deal of uncertainty in portrayal of the concepts. The capability of this abductive reasoning with uncertain causal relations of MOWL has a specific advantage over existing ontology languages when it comes to robust concept recognition.

Many of the dance steps are manifested through a sequence of postures. The individuality of the artistes brings in uncertainty in such sequences. It is possible to define such uncertain sequences with the spatio-temporal relations in MOWL. Further, many of the concepts in the domain, such as the mythical stories and the roles, are interrelated, implying that recognition of one leads to increased belief in the presence of another. The concept of *media property propagation* in MOWL facilitates such reasoning. Thus, the capabilities of MOWL in causal modeling, defining spatio-temporal compositions, and reasoning with media property propagation play an important role in recognizing the underlying concepts from audiovisual depictions of Indian classical dance forms. Such capabilities are not present in traditional ontology languages.

10.2.1   MOWL Language Constructs for Dance Ontology

In this section, we illustrate with examples how MOWL constructs are used for encoding the ICD ontology.

- **Encoding media properties**

  MOWL constructs are used to attach media observables like media examples and media features (MPEG-7 based features like color, edge histogram, composite media patterns like postures, human actions) to concepts in ICD ontology. Figure 10.2(a) shows how the concept Pranam is associated with its media properties.

- **Specifying property propagation rules that are unique to a media-based description of concepts**

  For hierarchy relations, media properties propagate from superclass to subclass and media examples propagate from subclass to superclass. There are relations in an ontology, different from hierarchy relations, which also permit the flow of media properties. For example, say *Odissi Dance is accompanied by Odissi Music*, then the former generally *inherits* the audio properties of the latter, though the relation *is accompanied by* does not imply a concept hierarchy [refer to figure 10.2(b)]. To allow specifications of such extended semantics MOWL declares several subclasses as detailed in chapter 5.

- **Specifying spatio-temporal relations**

  Many concepts observed in videos are characterized by specific spatial or temporal arrangements of component entities. The occurrence of such

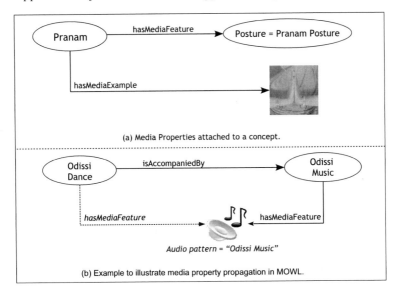

FIGURE 10.2: MOWL media property relations.

a relation is termed as an **event**, for example, a *dance step* in ICD. Classical dance sequences are choreographical sequences of some short dance steps, which are further choreographed as temporal sequences of classical dance postures performed by the dancer. We illustrate the spatio-temporal definition in MOWL of an ICD step labeled as *Pranam-DanceStep* in figure 10.3. It shows an observation graph for this concept. The smallest event unit is the temporal relation *followedBy*. An event entity node comprising `Folding of hands` *followedBy* the second event is created. The observation nodes connected to this event entity node are virtual evidence for the observation node `Folding of hands` and the second event entity node. Likewise for the second temporal relation *simultaneousWith* we have the two observation nodes: `Closing of eyes` and `Bowing of head`, and the *simultaneousWith* temporal relation. The observable media objects (images of postures) are also shown in the OM.

10.2.2 Multimedia Dance Ontology

To illustrate the idea of probabilistic reasoning, let us consider an Indian classical dance form *Odissi*, which is typically characterized by an opening act of *Mangalacharan* (invoking the gods). *Mangalacharan* is performed as a combination of three dance steps: *Bhumipranam* (salutation to the Earth), *Pushpanjali* (offering of flowers), and *TrikhandiPranam* (salutations). Each of these dance steps manifests in a series of postures. For example, *Bhumipranam*

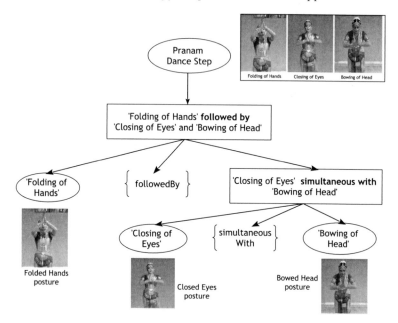

FIGURE 10.3: Spatio-temporal decomposition of `PranamDanceStep` (with permission of Vanshika Chawla [dancer]).

begins with a *Pranam* (salutation with folded hands), followed by a bending down with closed eyes. Each of these elementary postures can be detected using a trained set of classifiers. Thus, `OdissiDance`, `Mangalacharan`, and its constituent steps can be modeled as concepts, evidenced by the *observable* postures and their sequences, which can be modeled as media patterns. The concepts and their relationships are depicted in figure 10.4 and the corresponding MOWL encoding in figure 10.5.

Different artistes put different emphasis on these dance steps and bring in some variations in the postures, leading to uncertainties in these media manifestations. We associate CPTs with the media manifestations based on the significance associated with the postures in the literature. We depict a few other concepts in the figure, for example, `OdissiMusic`, which generally accompanies `OdissiDance`; and two of the well-known artistes, `Madhumita Raut` and `Yamini Krishnamurthy`, for the dance form, together with their possible media manifestations. These concepts are likely to co-occur with instances of sn Odissi dance recital, and hence, their corresponding media manifestations can also be expected with an Odissi dance performance. Note that some of the properties, such as the artistes or the postures, may not necessarily be exclusive for a particular dance form. In reality, the ontology is much larger and consists of several hundred nodes and relations. Some of these nodes are indicated with dotted lines and are not further elaborated in this chapter.

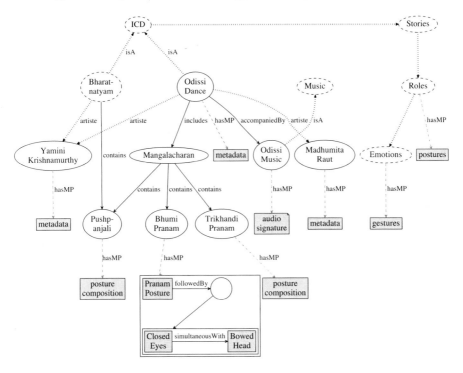

FIGURE 10.4: Part of the ICD multimedia ontology shown as a graph.

```
MOWL SNIPPET : ICD ONTOLOGY

1.  @prefix icd: <http://.../.../ICDOntology#>
2.  @prefix mowl: <http://.../.../mowl#>
3.
4.  icd:accompaniedWith a mowl:propagateMedia ;
5.      rdfs:Domain i:IndianClassicalDance;
6.      rdfs:Range i:IndianClassicalMusic .
7.  icd:artiste a mowl:propagateMedia ;
8.      rdfs:Domain icd:IndianClassicalDance ;
9.      rdfs:Range   icd:PerformingArtist .
10. icd:includes a mowl:propagateMedia ;
11.     rdfs:Domain icd:IndianClassicalDance ;
12.     rdfs:Range   icd:DancePiece .
13. icd:Music  a  owl:Class .
14. icd:OdissiMusic  a  icd:Music ;
15.     mowl:hasMediaPattern icd:audioSignature .
16. icd:IndianClassicalDance a  owl:Class .
17. icd:OdissiDance  a  icd:IndianClassicalDance ;
18.     icd:accompaniedWith  icd:OdissiMusic ;
19.     icd:includes icd:Mangalacharan ;
20.     icd:artiste icd:MadhumitaRaut ;
21.     icd:artiste icd:YaminiKrishnamurthy ;
22.     mowl:hasMediaObject  icd:metadata .
23. icd:Mangalacharan  a  icd:DancePiece ;
24.     mowl:hasMediaPattern   icd:BhumiPranam .
25. icd:followedBy a mowl:Predicate ;
26.     mowl:R_t  mowl:R00011 ; ...
27. icd:together  a mowl:Predicate ;
28.     mowl:R_t  mowl:R01110 ; ...
29. icd:PranamPosture a mowl:MediaPattern; ...
30. Icd:MadhumitaRaut  a  icd:Dancer ; ...
```

FIGURE 10.5: MOWL snippet of ICD ontology.

10.2.3   Concept Recognition and Semantic Annotation of Heritage Artifacts

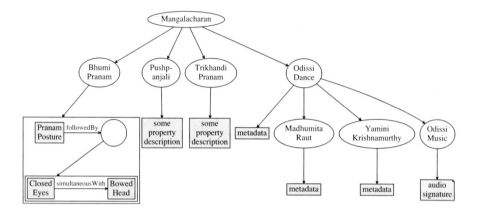

FIGURE 10.6: Observation Model for concept `Mangalacharan` of the ICD ontology.

The ontology can be used to create observation models for the different concepts in the domain following the algorithm given in section 5.8.1. The OM for the concept `Mangalacharan` (without CPTs) is shown in figure 10.6. Note that the concept is recognized with a multitude of evidence and failure of a feature detector has little impact on the overall performance. In particular, the belief in concept `Mangalacharan` is reinforced with establishment of the context `Odissi Dance` through recognition of related concepts, like the `Odissi Music` or either of the two artistes, because of media property propagation. Thus, the concept is likely to be recognized even if some of the constituent dance steps are not recognized, when the context is established. Further, the constituent elementary postures of complex concepts, like `Bhumipranam`, have more definitive features, and it is possible to build more accurate classifiers for them. Deployment of such pattern detectors and reasoning with their spatio-temporal composition improves the performance of detection of complex media properties.

Table 10.7 compares the concept recognition performance of the MOWL ontology-based approach with a traditional classifier-based approach. Approximately 36 percent of total entities in the ontology were selected for experimentation. The chosen entities represent a mix of elementary media patterns, complex media patterns, and abstract concepts. SVM-based classifiers were used for detecting elementary media patterns, such as the elementary dance postures and the genre of the music. Similar classifiers were used for the other experimental entities for the sake of comparison.

The results for some of the entities are shown in figure 10.7. As expected, the recognition performance in the MOWL-based approach remains the same

Entity	SVM		MOWL	
	Precision	Recall	Precision	Recall
Elementary media patterns				
ChawkPosture	0.84	0.87	0.86	0.88
PranamPosture	0.78	0.78	0.78	0.78
Complex media patterns				
FluteAction	1.00	0.50	0.95	0.89
BhumiPranam	0.73	0.85	0.86	0.89
Ardhacheera	0.91	0.83	0.90	1.00
Abstract Concepts				
KrishnaRole	0.70	0.60	0.84	0.89
Mangalacharan	0.47	0.47	0.84	0.84
OdissiDance	0.44	0.47	0.80	0.87
BattuDance	0.29	0.36	0.72	0.88
YashodaRole	0.67	0.55	0.78	0.85
MahabharatTheme	0.40	0.25	0.55	0.75
KrishnaTheme	0.70	0.32	0.75	0.83

FIGURE 10.7: Concept recognition results for the ICD ontology.

for low-level media patterns which are detected directly with specialized pattern detectors. The complex media patterns, such as the dance steps and actions, are described as temporal sequences of elementary postures in the MOWL-based approach. Superior results are achieved for these entities. The abstract concepts in the domain have no definitive features and it is difficult to design usable classifiers for them. The recognition performance with SVM-based classifiers is extremely poor for such entities. The MOWL-based approach that involves detecting elementary media patterns and combining the evidence provides significantly improved results in these cases.

10.2.4   Browsing and Querying Video Content

In this section, we show how the Indian classical dance ontology and video annotations of the ICD video, enable the creation of a semantic navigation environment in the cultural heritage repository. The semantic navigation is provided with an ontology-guided user interface for a video-browsing application.

The video database in the *Nrityakosha* system [129] consists of videos of Indian classical dance as well as their annotation files. These XML files for storing the conceptual annotations contain sufficient details to allow access to any section of the video: right from a scene, a shot, a set of frames depicting an event, a single frame denoting a concept, or an object in an audiovisual sequence. Each of these entities is labeled with one or more concepts in the domain ontology, through manual, supervised, and automatic annotation gen-

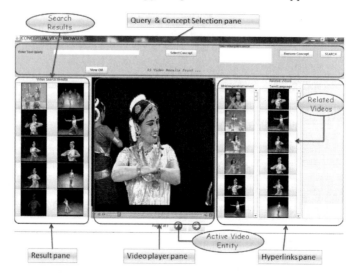

(a) Different panes of the browser

(b) Selection of concepts through a textual representation of the ontology

FIGURE 10.8: Conceptual video browser screenshots (with permission of Vanshika Chawla [dancer]).

eration, as detailed in section 7.5.2.3. The video annotation files in XML along with the actual videos and a MOWL ontology of the domain constitute the inputs to the video-browsing interface called Conceptual Video Browser (CVB), shown in figure 10.8. The domain ontology enriched with multimedia videos of the domain is parsed to produce a textual or graphical representation.

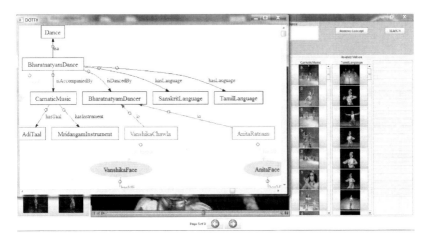

FIGURE 10.9: Conceptual video browser: Example I.

The user can view the domain concepts along with their relations and then select domain concepts to access. This browsing interface provides the user with a concept-based navigation of the video database, and she can easily browse and navigate progressively towards more relevant content. Using this interface, a user can retrieve video content from different videos, view annotations attached to a video entity and navigate to other similar video entities using hyperlinks. Instances of the same concept entity or related concept entities are hyperlinked together for this purpose. Example snapshots in figures 10.9 and 10.10 illustrate how CVB is used to browse the ICD video collection, with a MOWL ontology of the ICD domain being an input to the browsing tool. The system is able to show videos that are related to the search results on the basis of various ontological relations and thus provides a comprehensive browsing and navigational interface to the ICD collection.

In figure 10.9, the user wants to watch videos of a particular dance form and intends to view dance performances pertaining to a certain kind of beat and music. The user selects major dance form concept `BharatnatyamDance` from the ontology and submits a conceptual text query for search. Video search results contain thumbnails of Bharatnatyam dance videos. Related videos are shown in the hyperlink pane under two columns with labels *Carnatic Music* and *Tamil Language*. The user can click on the thumbnail of a video to select to play it. Further, the user selects to view the ontological relations of Bharatnatyam dance by pressing the "View OM" button. For further navigation, the user can select the the language *Tamil Language* or the music form *Carnatic Music* and browse videos pertaining to the selected concept.

In figure 10.10, the user chooses to see the OM for Carnatic music to navigate further. After viewing the ontological relations, he can select a musical

FIGURE 10.10: Conceptual video browser: Example I (continued) (With permission of Vanshika Chawla [dancer]).

beat (Taal), choosing - *Adi Taal*, which is related to *Carnatic Music*. This CVB screen shot shows the observation model for `CarnaticMusic` concept and its relations with other concepts, including the `AdiTaal` concept.

10.3   Intellectual Journey: Space-Time Traversal Using an Ontology

Heritage preservation requires preserving the tangibles (monuments, sculpture, coinage, etc.) and the intangibles (history, traditions, stories, dance, etc.). Besides these artifacts, there is a huge amount of background knowledge that correlates all these resources and establishes their context. This section discusses a new paradigm for heritage preservation — "an intellectual journey with space-time traversal" — which is more advanced than physical explorations of heritage sites and virtual explorations of monuments and museums. This paradigm proposes an experiential expedition into a historical era by using an ontology to interlink the digital heritage artifacts with their background knowledge. A multimedia ontology encoded in the Multimedia Web Ontology Language (MOWL) is used to illustrate this paradigm by correlating the digital artefacts with their history as well their living context in today's world. The user experience of this paradigm involves a virtual traversal of a heritage site or theme, with an ontology-guided navigation through space and time and a dynamic display of different kinds of media.

The *intellectual journey* proposed here is much more informative and advanced than physical exploration of heritage sites or the walk-through virtual tours currently offered by online museums and websites. Other related works

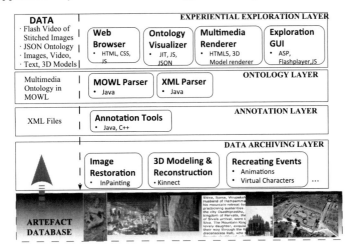

FIGURE 10.11: Software architecture of *intellectual journey* paradigm.

in cultural heritage preservation, like [102], make use of augmented and virtual reality but fail to present artifacts in context with background knowledge. A similar ontology-based framework for the digital exploration of cultural heritage objects has been used in [3], but the authors have not looked at traversal over time through correlation of classical and historical narratives with living traditions and folklore.

10.3.1 Ontology-based Interlinking of Digital Heritage Artifacts

Once a digital multimedia collection has been curated and annotated at different levels of granularity with semantic concepts of its domain, it is possible to provide semantic access to its constituent media files using a MOWL encoded domain ontology. As this representation allows collection-independent modeling of the domain, different kinds of media — images, text, video, graphics, audio — can be associated with the concepts, which aids in their perceptual modeling. This also helps in building semantic conceptual linkages between different modalities and can be used to provide interlinking of the digital artifacts that are part of a heritage collection. An algorithm for the same from [128] is shown in algorithm 9. This ontology-based interlinking of digital heritage artifacts is the basis of the experiential exploration of a heritage site or theme, with readily available tools to traverse in space and time and access e-heritage artifacts in context and on demand.

Algorithm 9 Multimedia retrieval through ontology-based interlinking

---

Inputs: a)Search term set $\mathcal{T}$ or clicked image $\mathcal{K}$; b) Bayesian Network Ω of the relevant MOWL ontology segment; c) Set $\mathcal{X}$ of MPEG-7 compatible XML files.

Output: Concept $\mathcal{C}$ and associated set $\mathcal{I}, \mathcal{V}, \mathcal{T}$ of images, videos, and text

1: Look up the mapping table which contains the mapping $\mathcal{K}, \mathcal{C}$ to get the concept $\mathcal{C}$ for image $\mathcal{K}$, and go to step 4.
2: Instantiate the leaf nodes in Ω, which match the search terms in $\mathcal{T}$.
3: Carry out Belief Propagation in the BN Ω.
4: Obtain set of concepts which have posterior probability $P(\mathcal{C}_i) >$ threshold. $\mathcal{C}$ is the concept with highest posterior.
5: **for** $i = 1$ to $|\mathcal{X}_r|$ **do**
6:     Search for the media segment descriptions with $\mathcal{C}$ as label.
7:     Add the corresponding media segment to set $\mathcal{I}$, $\mathcal{V}$, or $\mathcal{T}$ depending upon its media type
8: **end for**
9: Return the set $\mathcal{I}, \mathcal{V}, \mathcal{T}$

10.3.2 Ontology-based Framework for Space-Time Exploration

Figure 10.11 shows the software architecture of the framework that offers the intellectual exploration experience using an ontology. The different layers in this architecture are:

Digital Archiving Layer: This layer consists of all the technical modules for capturing and recreating the digital artifacts. The outputs of this layer are digital collections of the artifacts of different modalities, that is, image, text, 3D, video, and audio collections.

Annotation Layer: It consists of annotation tools required for annotating and labeling the documents in the artifact collections. This step assumes that a multimedia ontology for the domain has been acquired and semantic labels associated with domain concepts in the ontology are available for labeling the artifacts. Depending on the type of document (image, audio, video, text), different tools are required to identify artifacts within that document, for example, image cropping, selection, demarcating video segments, or identifying individual video frames. Semantic annotation involves labeling each artifact with one or more ontology concepts. This layer produces an MPEG-7-based XML file per document in the collections, with annotations for multiple artifacts in the document.

Ontology Layer: This layer consists of the language parsers for the ontology as well as for the XML files. The ontology is parsed and web-compatible visualization graph is produced. The XML files are parsed to produce sets/chains of artifacts that are linked through common ontology concepts. Thus the ontology not only correlates the concepts through domain knowledge, but also produces data linkages that serve the purpose of an ontology-based exploration and cross-modal access of media.

Experiential Exploration Layer: This layer is composed of web-based

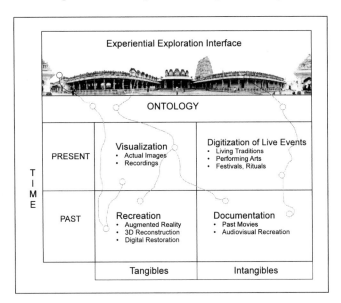

FIGURE 10.12: Ontology-based Space Time Exploration.

graphic user interfaces for exploring a heritage site, for ontology visualization, and for presenting or rendering the digital artifacts. The user interfaces with this layer to carry out an experiential exploration as proposed by our paradigm. A user of this system can explore a heritage site virtually, re-created using actual site images. Here are two examples of the different paths he can take in space and time with ontology-guided navigation. **Path1**: The user clicks on a temple icon on screen. It maps to a concept in the ontology. XML annotation files are searched for annotations pertaining to this concept. This search produces a set of images of the temple as it exists now, a text detailing its history, a video of a tour of its premises, and a 3D model of the temple as it was originally built. These are presented to the user. **Path2**: User views the video of a weekly market or bazaar (*present intangible*) as it happens at a temple site today. Then she selects a timeperiod (15th century) in history through the ontology. The system retrieves an animation movie (*past intangible*) that re-creates the bazaar as it used to happen in the 15th century in that temple. Figure 10.12 shows these paths as threads.

10.3.3 Intellectual Exploration of the Theme of `Girija Kalyana`

An ontology expert generates an ontology of the domain from a *tag dictionary*, which is created with domain knowledge derived from experts and other sources. This dictionary contains different tags that can be used for labeling different aspects of the artifacts. A sample of such a tag dictionary is shown in figure 10.13. The tag dictionary is used as the basis for providing tags for

- Location
  - Hampi
  - Lepakshi
- Temple
  - Virupaksha
  - Virbhadra
- Locale
  - PampaSarovar
  - ManmathaKonda
- Temple InnerLocation
  - KalyanaMandapa
  - GarbhaGriha
- Mural
  - GirijaKalyanaMural1
  - GirijaKalyanaMural2
- Scripture
  - ShivaPurana
- Empire
  - Vijayanagara

- Story
  - GirijaKalyana
- Characters
  - Hampamma
  - Pampapathi
  - Nandi
- Literature
  - Hariahara's GK
  - PampaMahatme
- Art Tradition
  - Pallavas
  - Cholas
- MuralContent
  - Fashion
  - BorderPattern
  - Floral
- Narratives
- Ritual
  - KalyanaUtsav
  - PhalPuje

FIGURE 10.13: Tag dictionary generated from annotation of `MuralPaintings`.

image segments, video shots, audio tracks, and text segments in the digital collection, which is the media data with which the heritage ontology is enriched. An example of tagging of a mural painting is shown in figure 10.14. Different annotation tools used for tagging different kinds of media produce XML annotation files in a standard format that is based on MPEG-7-based media descriptions. XML files produced from the annotation of artifacts contain semantic labels linked with media segment descriptions. Thus media segments with the same conceptual labels can be hyperlinked for access as required in retrieval through textual search queries or through a graphic user interface.

To demonstrate the intellectual journey paradigm using a MOWL ontology-based framework, the Indian heritage domain related to the UNESCO World Heritage Site of Hampi, in Karnataka, India, is chosen. The framework is generic and extensible to preserve any heritage site in the world for which background knowledge and heritage resources are available. The Hampi site is comprised of the ruins of Vijayanagara city, the former capital of the powerful Vijayanagara Empire, which flourished in South India during

MURAL PAINTINGS - HAMPI
Story: GIRIJA KALYANA –Marriage of Parvathi with Shiva
Location: Maharanga Mandapa ceiling, Virupaksha Temple, Hampi, Bellary District, Karnataka
Period: Between eighteenth – nineteenth centuries, post Vijayanagara (A.L. Dallapiccola 1997)
Purana: Shivapurana, Lingapurana

| *Upper level*: 3 Mythical crowned beings, with bird bodies in multi-lobed sub panels | Architectural elements with wedding guests, trees, animals & birds | One Mythical crowned creaturewith bird body on either side of the Kalyana Mandapa architectural elements with deities, trees, animals & birds | Architectural elements with wedding guests, trees, animals & birds | Two mythical crowned beings with bird bodies & one mythical being with bird face |

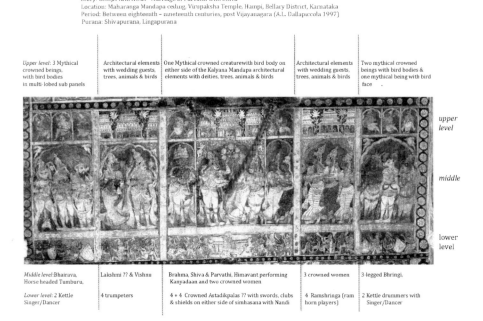

upper level

middle

lower level

| *Middle level*:Bhairava, Horse headed Tumburu, | Lakshmi ?? & Vishnu | Brahma, Shiva & Parvathi, Himavant performing Kanyadaan and two crowned women | 3 crowned women | 3-legged Bhringi, |
| *Lower level*: 2 Kettle Singer/Dancer | 4 trumpeters | 4 + 4 Crowned Astadikpalas ?? with swords, clubs & shields on either side of simhasana with Nandi | 4 Ramshringa (ram horn players) | 2 Kettle drummers with Singer/Dancer |

FIGURE 10.14: Tagging or annotation of a Girija Kalyana Mural (courtsey of the International Institute of Art, Culture and Democracy) (IIACD), Bangalore, India.

the 14th to 17th centuries. Several dynasties ruled during this period, patronized art and culture, built several new temples, and renovated and enlarged many old ones. Festivals and rituals involving community participation were encouraged, and in fact still take place at the site.

We focus on the theme of Girija Kalyana from our knowledge-base of the Hampi heritage. `Girija Kalyana` refers to the marriage between the goddess `Hampamma` (also known as Uma, Parvati, or Shakti) and the male Hindu god `Pampapathi` (also known as Shiva, Shankara, or Rudra). The concept manifests in both tangibles and intangibles of Hampi as well as some other Vijayanagara sites like Lepakshi in Andhra Pardesh, India. The tangible manifestations include mural paintings depicting the marriage on the walls of ancient temples in Hampi and Lepakshi, on stone and bronze sculptures depicting characters in the story, found in temples and museums, and so on. There is a story attached to the concept, the text of which is found in certain ancient scriptures like the `Shiva Purana`. Certain aspects of this mythological story are also found in narratives in old inscriptions, memorial stones and manuscripts. Narratives of some performing arts like folk dance, theater, and puppetry in areas around Hampi and Lepakshi abound in references to the marriage of Shiva and Parvati.

The Girija Kalyana context has a living heritage. Every year the marriage is celebrated in the ancient temple of Hampi. Communities from nearby areas congregate to attend the famous rituals of Kalyana Utsava (the wedding), preceeded by Phala puja (the engagement). Priests actually conduct a wedding of the deities in the temple, and people make offerings to please the gods. Traditional crafts like making puppets, toys, wooden door frames, and wall paintings are still being practiced in locations around Hampi. Many of these replicate images and patterns from the Girija Kalyana concept. It is a fashion in local communities to get *tattoos* depicting some Girija Kalyana patterns drawn on their person.

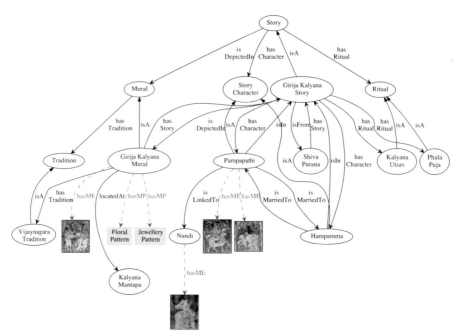

FIGURE 10.15: Indian heritage ontology snippet with a focus on the GirijaKalyana concept.

A multimedia ontology snippet of the Girija Kalyana context, after its creation from the tag dictionary in figure 10.13, is shown in figure 10.15. It shows how the story of the Girija Kalyana depicted by the concept Girija Kalyana Story in figure 10.16 is associated with the characters Hampamma and Pampapathi, with its depictions in mural paintings, rituals, and current traditions. The media properties attached to the concepts are shown with hasMediaPattern and hasMediaExample relations. Figure 10.16 shows the observation model generated for the concept `Girija Kalyana Story` with its media observables as well those of its related concepts.

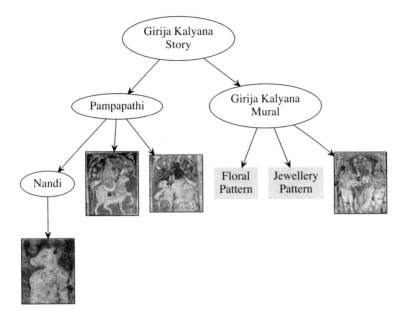

FIGURE 10.16: Observation model for the `GirijaKalyanaStory` concept.

10.3.4   Experiential Exploration Interface

Once the annotated repositories are available and the multimedia ontology has been generated, the user can do an experiential exploration of a heritage site or an abstract heritage theme using an interactive graphic user interface that has an ontology-guided navigation through space and time. The user can build dynamic narratives with the help of the ontology, and view a dynamically changing display of associated digital artifacts, in different kinds of media formats.

Figures 10.17 is a screen shot of the exploration GUI. The user experience of this journey involves watching videos, running image slide shows, and virtually exploring panorama views of stitched image sequences of various locations of the heritage site in the main panel. This is accompanied by a window view of the related ontology snippet as a frame of reference in the ontology panel. Navigation is ontology-guided where the sociocultural context of the frame is portrayed through a dynamic display of media and narratives in the data panel. While moving around virtually, the user has an option to click on certain visual cues in order to alter a visual display of image, text, and videos of the artifacts as well as change the perspective of the ontology visualization shown in a parallel window on screen. As the user context changes dynami-

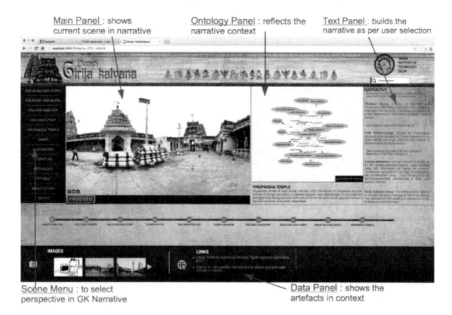

FIGURE 10.17: Screen shot with `VirupakshaTemple` concept selected.

cally, so does the display of artifacts and the view of the ontology. The user also has an option to navigate to a chosen concept in the ontology window by clicking on it. This too alters the set of artifacts displayed on screen. A text narrative is built in a separate frame called the text panel, based on the sequence of concept selections in the ontology, thus portraying a dynamically built story line as per the user's choice.

10.4   Cross-Modal Query and Retrieval

In this section, we discuss a framework to provide cross-modal semantic linkage between semantically annotated content of a repository of Indian mural paintings and a collection of labeled text documents of their narratives [130]. This framework, based on a multimedia ontology of the domain, helps preserve the cultural heritage encoded in these artifacts. The cross-modal query and retrieval application works on a collection of images depicting Indian mural paintings along with a set of text documents containing the narratives in these murals. Starting with a handcrafted basic ontology for the mural domain, we create a multimedia-enriched ontology by using a training set of labeled image segments from mural paintings and labeled text segments from the narratives. Using techniques of semantic matching between the text and image segments, we are able to build co-relations between the two through domain concepts of the ontology. Once an enriched ontology is available, it can be used to provide cross-modal access to mural image segments and text in the story documents. Besides this semantic correlation, the ontology also helps preserve the knowledge about mural paintings by catering to semantic queries pertaining to other aspects of the mural paintings such as painting style, location, time period, and so forth.

Indian temples have some of the largest murals in the world, painted across their ceilings and walls. The ceilings are replete with many stories from the epics and scriptures, though very few have survived the depredation of time. Each ceiling along the passageway of some of the temples is like a large panel narrating a story as one moves from the north to the south or east to west. The narrative in each panel flows from one scene to another, amid many intricate details, to follow the forms into a sequence of expressions by the characters. The imagery of the words from epic poems, the practices of rituals, the oral history, and the *Sthala puranas* (scriptures linked to a temple defining its architectural elements) remain a source of imagination and an inspiration for many of the temples. Each mural in the temple is further supported by a subtext and narration and sometimes even a painted manuscript.

10.4.1 Multimedia Ontology of Mural Paintings Domain

Multimedia ontology helps encode media properties of the semantic concepts. Here the MOWL ontology is used to associate mural image segments to ontology concepts. At the same time, text from narratives can also be associated with ontology segments as a media property. This helps build a semantic linkage between the two different modalities and can be used to provide cross-modal access to the heritage collections. The multimedia ontology of the mural domain is shown in figure 10.18.

The top layer of the ontology contains concepts like `Mural Painting`, `Temple`, `Mural Tradition`, `Mural Technique`, `Painting Style`, `Mural Content` and `Narrative`. The mural painting typically belongs to a temple, and therefore to a particular time period with which the temple is linked. It follows a certain mural tradition and therefore a typical painting style. It also has a narrative that further has a story. A story contains some characters. The mural also has some other content in addition to the depiction of the story or narrative. It has some border patterns like floral or geometric patterns, and some fashion patterns like hairstyles and jewelery. These media patterns definitely portray the history of the time period to which the mural belongs. The artistic style is also visible in the way characters are painted and placed in the mural. Multimedia ontology can be used to associate these visible patterns with concepts, and thus encode all this background knowledge very effectively. In fact, some of the concepts have mural image segments attached to them through the *hasME* or *hasMediaExample* relation.

There are hierarchical relations between subclasses and superclasses as well as between instances and the parent class. Other relations that are non hierar-

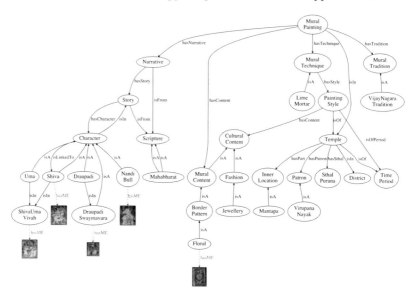

FIGURE 10.18: Graphical view of the multimedia ontology of Indian mural Paintings domain.

chical are such that they propagate some media properties from the *domain* of the relation to the *range*. An example of this is the relation *is In* between the character `Shiva` and the story `Shiva Uma Vivaha`. The media example image of the story `Shiva Uma Vivaha`, which shows `Shiva` getting married to `Uma`, and so has an image of `Shiva`, is thus also a media example of the concept `Shiva`. This kind of semantics is not provided by a standard ontology language, and thus makes MOWL the ideal representation to encode such knowledge.

10.4.2   Ontology-based Cross-Modal Retrieval and Semantic Access

This section shows how the multimedia ontology, by its collection-independent modeling of the domain, provides a robust basis for cross-modal access to knowledge of the domain collected in repositories stored in varied media formats. The heritage collection comprises of digitized images of mural paintings from the Vijayanagara period in Indian history. The other repository is of stories from Indian mythology and folklore, with three different versions of each story on an average. Text segments from the stories are labeled and tagged for different characters, episodes, locations, and so on.

10.4.2.1 LDA-based Image and Text Modeling

Latent Dirichlet Allocation (LDA) [19] based topic modeling is used to model the image and text segments in the heritage repositories. For the collection of text narratives, segments based on the tagging of the text are generated. Each of these text segments is first pre-processed (removal of stop words, stemming, etc.) to get the desired text data set before estimating with GibbsLDA++. This preprocessed data set is then fed to LDA. The mural images are broken down into segments based on their labeling by the experts. A bag of SIFT [124] descriptors is extracted from each image segment, and the SIFT descriptors extracted from each image are vector quantized with a codebook, producing a vector of visual word counts per image. Besides this *bag of words* representation, a lower-dimensional representation for images is generated, similar to that for text, by fitting an LDA model to visual word histograms and representing images as a distribution over topics.

10.4.2.2 Semantic Matching

In the mural paintings domain, the desired cross-modal retrieval for a text or image query is to show a story or narrative linked to a mural selected by the user; show image segments from different mural paintings that depict the narrative matching the text query; retrieve mural image segments that depict a particular character from a narrative; and so on. For example, a user may query for mural segments depicting `Shiva`, and when those segments are displayed, one of which might be the mural segment depicting `ShivaUmaVivaha`, the marriage of `Shiva` and his consort `Uma`, then the user might ask to see the narrative behind the story, and so on.

Let us consider the problem of information retrieval from a database $B = I_1, ..., I_B$ of semantically labeled image documents and from a database $N = T_1, ..., T_N$ of semantically annotated text segments. The semantic labels in each repository correspond to semantic concepts in the linked ontology. Images and text are represented as vectors in feature spaces R^I and R^T respectively. As illustrated in figure 10.19, the ontology helps to establish a one-to-one mapping between points in R^I and R^T. The goal of cross-modal retrieval is to return the closest match to a text query $T_q \in R^T$ in the image space R^I, and the same for an image query ($I_q \in R^I$) in the text space R^T.

Semantic matching means to map images and text to representations at a higher level of abstraction, where a natural correspondence can be established. This is obtained by augmenting the database B with a vocabulary $L = 1...L$ of semantic concepts from the domain ontology. Two mappings Π^I and Π^T are then implemented using classifiers of text and images, respectively. Π^T maps a text $T \in R^T$ into a vector Π^T of posterior probabilities $P_{W|T}(w|T)$; where $w \in 1...L$ with respect to each of the classes in L. The space S^T of these vectors is referred to as the semantic space for text, and the probabilities $P_{W|}(w|T)$ as semantic text features. Similarly Π^I maps an image I into a vector Π^I of

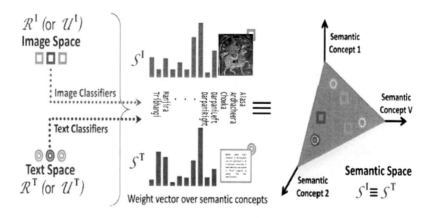

FIGURE 10.19: Semantic matching maps text and images into a semantic space.

semantic image features $P_{W|I}(w|I)$; where $w \in 1...L$ in a semantic space for images S^I.

10.4.2.3 Cross-Modal Retrieval

Depending on the precise nature of the probability model, naïve Bayes classifiers can be trained very efficiently in a supervised learning setting. The basic idea of this model is that each category has its own distribution over the codebooks, and that the distributions of each category are observably different. Given a collection of training examples, the classifier learns different distributions for different categories. For both the modalities in our work, posterior probabilities are computed that refer to the semantic concept of the ontology, and thus images and text are mapped to the same semantic space. Given any test set from any of the modalities, it is classified to one of the semantic concepts. KL (Kullback-Leibler) divergence is used to measure the distance between probability distribution of each document with all other documents for both the modalities. Given a query image I, represented by $\Pi_I \in S^I$, cross-modal retrieval will find the text T, represented by $\Pi_T \in S^T$ that minimizes $D_{KL}(I,T) = d(\Pi_I, \Pi_T)$ for some suitable distance measure d between probability distributions. Distance measure considered here is KL (Kullback-Leibler) distance. An illustration of cross-modal retrieval using semantic matching and search results using KL divergence after the test set has been classified by Bayesian classifier is shown in figure 10.20.

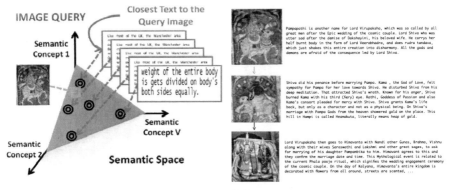

(a) Text retrieved given an image query.

(b) Search results using naïve Bayes classifier and KL distance.

FIGURE 10.20: Cross-modal retrieval using semantic matching

10.5 Conclusion

In this chapter, we illustrated the application of multimedia ontology and its representation through the Multimedia Web Ontology Language in preserving tangible and intangible heritage through improved access of digitized heritage artifacts. The MOWL ontology-based framework provides a conceptual linkage between the heritage resources at the knowledge level and multimedia data at the feature level. This generic technology using a multimedia ontology and a MOWL-based framework can be utilized in any multimedia application for heritage knowledge management. In this chapter, the advantage of using this technology has been illustrated through some applications in the domain of Indian cultural heritage preservation. These applications include a system to preserve the intangible heritage of Indian classical dance, which provides semantic annotation of, and access to, the audiovisual data recordings of dance performances. Another application discussed is an intellectual exploration framework based on an ontology that allows space-time traversal with the help of ontology-based interlinking of digital artifacts. Cross-domain retrieval from digital collections of Indian mural paintings and their stories and narratives in text is another illustrative example.

Chapter 11

Open Problems and Future Directions

11.1 Knowledge and Human Society

Progress of human civilization has always been driven by knowledge acquired and curated over generations. Philosophers in the 17th century realized that humans being the creator of their own society, can understand its dynamics and change it for the betterment of the humankind. Knowledge about natural phenomena, or "scientific" knowledge, has been systematically applied for progressive material benefit of human beings over the last few centuries. Intellectual growth and sustainable development has been fueled by knowledge about the cultural values of the societies. In modern times, digital computers and worldwide networking present an opportunity to digitize and share information pertaining to all walks of human life. The "digital revolution" had led to the availability of a mind-boggling volume of information online. For effective utilization of such vast amounts of information, knowledge management for culture and development is considered extremely important by many countries and has become a significant thrust of the United Nations as well. Human civilization is poised to transit from an information-driven society to a knowledge-driven society.

11.2 Knowledge Organization and Formal Representation

The knowledge possessed by human society today has accumulated over generations. It represents the efforts by numerous scholars expanding the frontiers of knowledge one step at a time, as well as common knowledge acquired from day-to-day observations and applications. The advancement of knowledge generally integrates and extends prior knowledge at any point of time. For example, Newton's contributions to physical laws are extensions to the theories proposed by Kepler and Galileo. At the same time, prior knowledge has often been challenged, refined, discarded, and sometimes rediscovered in the pursuit of knowledge. Examples include refinement of Newton's laws of motion with Heisenberg's uncertainty principles and rediscovery of the values of some traditional medical practices in recent times.

In ancient times, knowledge was traditionally transmitted from one generation to another through verbal communication. In later times, it has been documented in different forms, such as cave paintings, carvings on stone plaques and manuscripts. Invention of the printing press set the stage for dissemination of knowledge to the masses. The proliferation of web technologies facilitates contribution by any individual to the global pool of human knowledge, exemplified by collaboratively created knowledge resources, like Wikipedia. The advent of web technology has a multiplier effect on the accumulation in the human knowledge pool.

Knowledge generally exists in unstructured, distributed and unorganized form, and its bits and pieces are often interlinked. Information scientists have tried to organize human knowledge following natural order and using various classification and facet analysis methods [28]. The knowledge domain has been partitioned and subpartitioned to various subject headings, which are in turn cross-referenced to establish their linkages. Documents have been classified using such classification schemes, physically organized on the shelves of the libraries, and catalogs have been created for easier access.

In modern times, the science of artificial intelligence attempts to utilize collective human knowledge in specific domains in realizing computerized systems that can perform *intelligent* functions. This requires a formal representation of human knowledge that is machine processable. This has led to the development of several knowledge representation techniques during the last decades of the 20th century. The advancement in networking technology provides an opportunity for interaction between, and integration of, knowledge stored at the various network nodes. The proponents of Semantic Web technologies envision a world where the machines with adequate horsepower will be able to interpret and collaboratively act upon the information spread across the Internet to produce useful results for the benefit of human beings. The World Wide Web Consortium (W3C) works with the objective to standardize information and knowledge representation schemes on the web to facilitate such integration. These efforts have resulted in the development of a suite of standardized languages, such as HTML, XML, RDF, and OWL, each ensuring interoperability at a syntactic or semantic layer. The evolution of Semantic Web technologies is discussed in detail in chapter 2 of this book.

11.3   The Perceptual World and MOWL

The formal knowledge representation schemes developed in the wake of the Semantic Web movement dealt with symbolic knowledge and textual information, which had been prevalent on the network at that time. They have been used in many applications, such as creation of structured information like the *infobox* from unstructured information sources [185] and intelligent travel agents [30] that can integrate information from multiple heterogeneous resources.

The current century has witnessed progressive proliferation of information in multimedia form on the Internet. Multimedia information not only provides multisensory stimulation to the audience but is often the only or the preferred means to convey some information. For example, the position and shape of a brain tumor can be best illustrated by an image, rather than a textual description, while the latter may be a more convenient tool to express diagnosis and prognosis. The experience of a classical dance performance through a few multimedia examples cannot be replicated by any amount of textual description. Machine interpretation and utilization of such information in conjunction with knowledge represented in symbolic form needs additional knowledge about the perceptual properties of concepts. Such knowledge has been available to human society either in tacit form or in the form of informal documents. There has been no formal representation scheme for encoding such perceptual knowledge. Our proposal of a multimedia knowledge representation scheme (MOWL) in this book is an attempt to bridge the gap.

We have argued that knowledge representation for media properties of concepts requires a fundamentally different approach from the symbolic knowledge representation. The difference originates from the perceptual nature of multimedia artifacts as opposed to the conceptual nature of symbolic representations, such as text. One of our key contributions in this research monograph is to propose an *observation model* that is a tangible media-based description of abstract concepts. The knowledge representation scheme proposed by us can encode media property descriptions of concepts in addition to establishing domain-specific relations between them. Thus, it can create observation models for concepts incorporating contextual cues in a domain context. The probabilistic reasoning scheme supported with MOWL enables applications to cope with the inherent variabilities in media depictions of concepts while interpreting multimedia contents. The use of the knowledge representation and reasoning scheme to solve several real-world problems, involving retrieval, recommendation, and integration of multimedia contents from distributed collections, has been presented in different chapters of this book.

11.4 The Big Debate: Data-Driven *vs.* Knowledge-Driven Approaches

Proliferation of a huge volume of annotated and unannotated multimedia data on the Internet and the availability of massive amounts of computing power make statistical data-driven interpretation of media data an attractive proposition [119]. It can be disputed if ontology has a place in resolving multimedia semantics in the data-driven world. The experiential nature of the interpretation of multimedia data lies at the center of controversy. There are proponents and opponents to the dispute [168]. A primary argument of the opponents is that an ontology assumes an interpretation of the contents independent of the interpreter and his past experience, which is crucial to perception of multimedia data. To illustrate the point with a simple example, the interpretation of a video of a racing car by an automobile enthusiast generally differs from that of a naive viewer. Thus, an ontology *forces* its creator's views rather than those of the person experiencing the multimedia content. This justifies the use of in-context personalized machine-learning techniques for media interpretation. Examples of such machine-learning techniques using media features and annotations for visual and audio data have been presented in [221, 98].

The same argument can be used in a different way by its proponents. An ontology representation scheme provides a mechanism to encode human knowledge curated over generations. The prior ontology representation schemes dealt with conceptual entities and their relations. MOWL extends the capability to encode perceptual media properties of the entities. Thus, an observation model created from a well-curated conceptualized knowledge can be a reflection of human knowledge for media-linked phenomena. Thus, an ontology based interpretation of media artifacts is an outcome the collective experience of human society accumulated over generations that may be preferred over personal and uninformed interpretation in several complex domains, such as architecture, visual and performing arts, and heritage collections, just to name a few. Further, an ontology, being a domain model, can correlate media contents with other sources of information. It can thereby discover additional information about the media contents that are not present in the media instance itself.

11.5   Looking Forward: Convergence between Knowledge-Based and Data-Driven Approaches

Transforming human wisdom to a formal ontology representation is a challenge. Several methods to build ontologies from formal schema descriptions and Unified Modeling Language (UML) representations have been proposed in the literature. Another way to build such an ontology is by parsing the digitized versions of traditional knowledge sources, such as books and manuscripts [92]. The illustrations in these documents can serve as media examples for the concepts. Other approaches for creating ontology include formal methods for eliciting tacit information from the experts, formal encoding, and validation [147, 14], similar to the the requirement gathering process in traditional information system design. No method has so far been proposed for capturing media-based descriptions of complex phenomena to a formal knowledge representation scheme, such as MOWL. We have mostly assumed an informal mechanism of eliciting expert knowledge in this book.

In chapter 7, we indicated how segments of a MOWL ontology can be learned through statistical analysis of annotated multimedia data. The current technologies for data-driven approach for example, *deep learning* can be applied to learn the discriminatory features of "tangible" concepts that manifest into distinct albeit diverse media features. In an observation model, these concept nodes are directly connected to the media property description nodes. It can be argued that a complex media event can be inferred by recognizing its individual constituents. However, the gestalt model of psychology asserts that perception in the human mind is guided by observation of the whole phenomenon before or together with perception of its constituent parts. This is evidenced by recognition of street scenes in a blurry image, even when the constituent objects, such as the buildings, pedestrians, and motor-cars, cannot be identified in isolation. Similarly, a speech segment can be understood even when some words get muffled. This is possible because of the capability of the human mind to re-create the missing parts and reconstruct the whole by virtue of its past experience. This points to the necessity for creating holistic observation models for high-level concepts for their recognition. For example, the observation model of a classical dance should comprise not only the constituent dance steps but also accompanying music form, identity of the possible exponents in the dance form, and other such information. Knowledge representation with MOWL provides a mechanism for creating such complex schema descriptions for high-level concepts with collective human knowledge. Creation of such schema for such complex media phenomenon without using human knowledge is beyond the realm of current technology.

Thus, our contention is that both formal representation of human knowledge and machine-learning have their own places in solving real-life problems. We further contend that an ontology representation should not be static but should get dynamically updated with new facts to represent advancement of knowledge. Like in human society, the evolutionary knowledge base need not be monolithic and confined to a single node but may be distributed over the network. New knowledge resources may either complement or replace older resources in such dynamic scenario. In this context, alignment of distributed knowledge resources and its use in processing huge volumes of information distributed over the network, as described in chapter 9, assumes special significance.

To summarize, we see that an ontology representation scheme of MOWL provides a mechanism for encoding the media properties of concepts and to reason with them in a domain context. The ontology can be created using different means, such as by eliciting knowledge from the domain experts and by analyzing traditional knowledge resources. The reasoning scheme in MOWL can be used for constructing holistic observation models for concepts that manifest as complex media phenomena. The observation models can be used for robust recognition of abstract concepts, which is not possible with other techniques. Machine-learning principles can be used in conjunction with MOWL to learn the media properties of low-level tangible concepts encoded in an ontology. MOWL can be used for information integration from heterogeneous multimedia repositories distributed over the web. It is possible for fragments of MOWL ontology to be aligned and to be used in conjunction with each other to solve real-life problems. Thus, we contend that MOWL has the potential to become the foundation for the *semantic multimedia web* that will prevail in the future.

Appendix A

MOWL Schema in Notation3

```
@prefix rdf: <http://www.w3.org/1999/02/22-rdf-syntax-ns#> .
@prefix rdfs: <http://www.w3.org/2000/01/rdf-schema#> .
@prefix xsd: <http://www.w3.org/2001/XMLSchema#> .
@prefix owl: <http://www.w3.org/TR/owl-ref#> .
@prefix mowlDatatypes: <file:///MOWL/mowlDatatypes.xsd#> .
@prefix mowl: <file:///MOWL/mowl.owl#> .

mowl:Concept a owl:Class ;
rdfs:label "Concept" ;
rdfs:comment "A MOWL class, to encode concepts which represent
  the real world objects or events." .

mowl:MediaPattern a owl:Class ;
rdfs:label "MediaObject" ;
rdfs:comment "A MOWL class, to represent media observable
  concepts i.e. manifestation of a concept in media. This class
  is defined to allow association of media properties with
  entities in MOWL." .

mowl:MediaExample a owl:Class ;
rdfs:label "Media Example" ;
rdfs:comment "This class is for multimedia Examples (images,
  videos, audio files,text files etc.) associated with a
  concept." .

mowl:hasMediaPattern a owl:ObjectProperty ;
rdfs:domain mowl:Concept ;
rdfs:range mowl:MediaPattern .

mowl:hasMediaExample a owl:ObjectProperty ;
rdfs:domain mowl:Concept ;
rdfs:range mowl:MediaExample .

mowl:hasURI a owl:DatatypeProperty ;
rdfs:comment "Property to associate a URI with Media examples
  and media patterns for their location." ;
```

```
rdfs:domain (owl:unionOf (mowl:MediaPattern mowl:mediaExample));
rdfs:range xsd:anyURI .

mowl:propagateMedia a owl:ObjectProperty ;
rdfs:comment "An Object Property, different from Inheritance that
  allows propagation of properties between concepts, i.e. the
  media properties flow from the range to domain." ;
rdfs:domain mowl:Concept ;
rdfs:range mowl:Concept .

mowl:ComplexConcept a owl:Class ;
rdfs:comment "A sub-class of Concept class to represent compo-
  sition of concepts related via spatio-temporal relations." ;
rdfs:subClassOf mowl:Concept ;
  rdfs:subClassOf [
            a owl:Restriction ;
            owl:onProperty mowl:Subject ;
            owl:Cardinality "1"^^xsd:nonNegativeInteger
        ] ;
    rdfs:subClassOf [
            a owl:Restriction ;
            owl:onProperty mowl:hasPredicate ;
            owl:Cardinality "1"^^xsd:nonNegativeInteger
        ] ;
    rdfs:subClassOf [
            a owl:Restriction ;
            owl:onProperty mowl:Object ;
            owl:Cardinality "1"^^xsd:nonNegativeInteger
        ] .

mowl:Predicate a owl:Class ;
rdfs:comment "Spatio-temporal predicates encode the various
  spatio-temporal and temporal relations between concepts" .

mowl:Subject a owl:ObjectProperty ;
rdfs:domain mowl:Concept ;
rdfs:range mowl:Concept" .

mowl:Object a owl:ObjectProperty ;
rdfs:domain mowl:Concept ;
rdfs:range mowl:Concept" .

mowl:hasPredicate a owl:ObjectProperty ;
rdfs:domain mowl:ComplexConcept ;
rdfs:range mowl:Predicate" .
```

```
mowl:FuzzyAllen a owl:Class ;
owl:oneOf ( mowl:R10000 mowl:R11000 mowl:R10000 mowl:R11000
 mowl:R00011 mowl:R00001 ... mowl:precedes mowl:meets
 mowl:metBy mowl:precededBy" ... ) .

mowl:R10000 a mowl:FuzzyAllen .
mowl:R11000 a mowl:FuzzyAllen .
...
mowl:R00011 a mowl:FuzzyAllen .
mowl:R00001 a mowl:FuzzyAllen .
mowl:precedes a mowl:FuzzyAllen ;
owl:sameAs mowl:R10000 .


mowl:FuzzyContainment a owl:Class ;
owl:oneOf ( mowl:R0110 mowl:R1100 mowl:R1011 mowl:R0111 mowl:R1111
  mowl:R1110 mowl:R0110 mowl:outside mowl:contains mowl:inside
  mowl:overlaps mowl:touching mowl:skirting ) ;
rdfs:subClassOf [
            a owl:Restriction ;
            owl:onProperty mowl:hasP ;
            owl:Cardinality "1"^^xsd:nonNegativeInteger
        ] ;
   rdfs:subClassOf [
            a owl:Restriction ;
            owl:onProperty mowl:hasQ ;
            owl:Cardinality "1"^^xsd:nonNegativeInteger
        ] ;
   rdfs:subClassOf [
            a owl:Restriction ;
            owl:onProperty mowl:hasR ;
            owl:Cardinality "1"^^xsd:nonNegativeInteger
        ] ;
rdfs:subClassOf [
            a owl:Restriction ;
            owl:onProperty mowl:hasS ;
            owl:Cardinality "1"^^xsd:nonNegativeInteger
        ] ;

mowl:RX a owl:objectProperty ;
rdfs:domain mowl:Predicate ;
rdfs:range mowl:FuzzyAllen .

mowl:RY a owl:objectProperty ;
rdfs:domain mowl:Predicate ;
```

```
rdfs:range mowl:FuzzyAllen .

mowl:RZ a owl:objectProperty ;
rdfs:domain mowl:Predicate ;
rdfs:range mowl:FuzzyAllen .

mowl:RT a owl:objectProperty ;
rdfs:domain mowl:Predicate ;
rdfs:range mowl:FuzzyAllen .

mowl:RC a owl:objectProperty ;
rdfs:domain mowl:Predicate ;
rdfs:range mowl:FuzzyContainment .

mowlDatatypes:probValType a rdfs:Datatype ;
rdfs:isDefinedBy <file:///MOWL/mowlDatatypes.xsd> ;
rdfs:comment "type to encode decimal values from zero to one" .

mowlDatatypes:axisPtType a rdfs:Datatype ;
rdfs:isDefinedBy <file:///MOWL/mowlDatatypes.xsd> ;
rdfs:comment "string or decimal type to accommodate "-infinity",
        "+infinity" as well as other decimal points on the axis." .

mowlDatatypes:tupleType a rdfs:Datatype ;
rdfs:isDefinedBy <file:///MOWL/mowlDatatypes.xsd> ;
rdfs:comment "tuple (t,v) type i.e. a pair indicating a point
        t on the axis with the fuzzy membership value v" .

mowlDatatypes:FuzzyVector a rdfs:Datatype ;
rdfs:isDefinedBy <file:///MOWL/mowlDatatypes.xsd> ;
rdfs:comment "Fuzzy vector of many tuples" .

mowl:hasFuzzyAllenVector a owl:DatatypeProperty ;
rdfs:domain mowl:FuzzyAllen ;
rdfs:range mowlDatatypes:FuzzyVector .

mowl:hasT a owl:DatatypeProperty ;
rdfs:subPropertyOf mowl:hasFuzzyAllenVector .

mowl:hasU a owl:DatatypeProperty ;
rdfs:subPropertyOf mowl:hasFuzzyAllenVector .

mowl:hasV a owl:DatatypeProperty ;
rdfs:subPropertyOf mowl:hasFuzzyAllenVector .
```

```
mowl:hasW a owl:DatatypeProperty ;
rdfs:subPropertyOf mowl:hasFuzzyAllenVector .

mowl:hasX a owl:DatatypeProperty ;
rdfs:subPropertyOf mowl:hasFuzzyAllenVector .

mowl:hasFuzzyContainmentVector a owl:DatatypeProperty ;
rdfs:domain mowl:FuzzyContainment ;
rdfs:range mowlDatatypes:FuzzyVector .

mowl:hasP a owl:DatatypeProperty ;
rdfs:subPropertyOf mowl:hasFuzzyContainmentVector .

mowl:hasQ a owl:DatatypeProperty ;
rdfs:subPropertyOf mowl:hasFuzzyContainmentVector .

mowl:hasR a owl:DatatypeProperty ;
rdfs:subPropertyOf mowl:hasFuzzyContainmentVector .

mowl:hasS a owl:DatatypeProperty ;
rdfs:subPropertyOf mowl:hasFuzzyContainmentVector .

<!-- Constructs for Uncertainty Specification -->
mowl:hasCPT a owl:ObjectProperty ;
rdfs:comment "Property to associate a CPT with a Concept.
  If the concept has multiple parents which are independent,
  then there is one CPT per parent." ;
rdfs:domain mowl:Concept ;
rdfs:range mowl:CPTable .

mowl:CPTable a owl:Class ;
rdf:label "Conditional Probability Table" ;
rdf:comment "Class to encode the uncertainty and encapsulate
  the CPTs" ;
rdfs:subClassOf [
a owl:Restriction ;
owl:onProperty mowl:conditionedOn ;
owl:minCardinality "1"^^xsd:nonNegativeInteger
] ;

rdfs:subClassOf [
a owl:Restriction ;
owl:onProperty mowl:hasCPRow ;
owl:minCardinality "1"^^xsd:nonNegativeInteger
] .
```

```
mowl:conditionedOn a owl:ObjectProperty ;
rdfs:comment "Specifies the parent/s of a concept conditioned
  on which the CPT entries are based." ;
rdfs:domain mowl:CPTable ;
rdfs:range mowl:Concept .

mowl:hasRow a owl:ObjectProperty ;
rdfs:domain mowl:CPTable ;
rdfs:range mowl:CPRow .

mowl:CPRow a owl:Class ;
rdfs:comment "Conditional Probability row - a row of the CPT." .

mowl:parentStates a owl:DatatypeProperty ;
rdfs:comment "Specifies the vector of Parent States in a string.
  Mostly there is one parent, and the state is either 1 or 0,
  meaning concept is true or false i.e. present or absent." ;
rdfs:domain mowl:CPRow ;
rdfs:range xsd:String .

mowl:probValues a owl:DatatypeProperty ;
rdfs:comment "Specifies the Probability values for each of the
      two states of the concept, when the parent variables are in
      the state as specified. The value is a decimal number in the
      range [0,1]." ;
rdfs:domain mowl:CPRow ;
rdfs:range mowlDatatypes:probabilityVector .
```

Appendix B

MOWL Syntax Structural Specification

A structural specification of MOWL provides the foundation for the implementation of MOWL tools such as parsers and reasoners. It describes the conceptual structure of MOWL ontologies and thus provides an abstract representation for all syntaxes of MOWL.

This structural syntax specification is an extension of general definitions in OWL2 syntax specification [21]. A MOWL ontology consists of the following three different syntactic categories :

1. Entities - Classes, Properties (Object and Data), Individuals, which are identified by Internationalized Resource Identifiers (IRIs). They form the primitive terms and constitute the basic elements of an ontology.

2. Expressions — complex notions in the domain.

3. Axioms — statements that are asserted to be true in the domain

These three are used to express the logical part of MOWL ontologies. They are interpreted under a precisely defined semantics that allows useful inferences to be drawn. Entities — Class represents sets of individuals, including media properties; datatypes are sets of literals such as strings, integers, or more complex user-defined XML Schema Deficition (XSD) and Web Service Definition Language (WSDL) datatypes; object and data properties can be used to represent relationships in the domain. These properties include some new kinds of relations like media associations, media propagation, spatio-temporal relations, and uncertainty specifications.

MOWL specification here is written using the Backus-Naur From (BNF) notation, summarized in table 1 in OWL2 syntax. In this document, examples assume the following name-space prefix bindings unless otherwise stated :

Prefix	IRI
rdf	http://www.w3.org/1999/02/22-rdf-syntax-ns
rdfs	http://www.w3.org/2000/01/rdf-schema
xsd	http://www.w3.org/2001/XMLSchema
owl	http://www.w3.org/2002/07/owl

OntologyDocument := { prefixDeclaration } MowlOntology
prefixDeclaration := 'Prefix' '(' prefixName '=' fullIRI ')'

MowlOntology := 'MowlOntology' '(' [MowlOntologyIRI [versionIRI]]
                directlyImportsDocuments
                OntologyAnnotations
                axioms ')'

axioms := Axiom
...
// MOWL datatypes derived from standard XSD datatypes //
mowl:probValType
mowl:axisPtType
mowl:tupleType
mowl:FuzzyVector
mowl:probabilityVector

// Extending Classes of OWL2 with MOWL Classes //

// Basic Classes of MOWL //
Concept := IRI
// Media based classes of MOWL //
MediaExample := IRI
MediaPattern := IRI
// Complex Concept classes to represent Spatio-temporal constructs //
ComplexConcept := IRI
Predicate := IRI
FuzzyAllen := IRI
FuzzyContainment := IRI
// Uncertainty specification classes of MOWL //
CPTable := IRI
CPRow := IRI
...
// Extending OWL2 Expressions with MOWL Class Expressions //
ClassExpression := ... | BasicClassExpression | MediaClassExpression
| ComplexClassExpression | CPTClassExpression
BasicClassExpression := 'Concept' | 'Concept'
MediaClassExpression := 'MediaExample' | 'MediaPattern'
ComplexClassExpression := 'ComplexConcept' | 'Predicate' | 'FuzzyAllen'
| 'FuzzyContainment'
CPTClassExpression := 'CPTable' | 'CPRow'

// Extending Object Properties of OWL2 with MOWL Object Properties //
ObjectPropertyExpression := ObjectProperty | MediaPropertyExpression
| STPropertyExpression | AllenPropertyExpression | ContainmentProper-
tyExpression | CPTPropertyExpression

// *MOWL Object Properties* //
MediaPropertyExpression := 'hasMediaExample' | 'hasMediaPattern'
| 'propagateMedia'
STPropertyExpression := 'hasPredicate' | ComponentPropertyExpression
ComponentPropertyExpression := 'Subject' | 'Object'
AllenPropertyExpression := 'RX' | 'RY' | 'RZ' | 'RT'
ContainmentPropertyExpression := 'RC'
CPTPropertyExpression := 'hasCPT' | 'hasCPRow' | 'conditionedOn'
...
// *Extending Data Properties of OWL2 with MOWL Data Properties* //
DataPropertyExpression := DataProperty | MediaDataExpression | Allen-
DataExpression | ContainDataExpression | CPTDataExpression
...
// *MOWL Data Properties* //
MediaDataExpression := 'hasURI' | 'usesExample'
AllenDataExpression := 'has_T' | 'has_U' | 'has_V' | 'has_W' | 'has_X'
ContainDataExpression := | 'has_P' | 'has_Q' | 'has_R' | 'has_S'
CPTDataExpression := 'parentStates' | 'probValues'

// *Extending OWL2 axioms by MOWL axioms* //
axiom := ... | ClassAxiom | MowlMediaAxiom | MowlSTAxiom | Mowl-
CPTAxiom
...
// *MOWL hierarchical relation is same as OWL* //
ClassAxiom := SubClassOf | ...
SubClassOf := 'SubClassOf' '(' axiomAnnotations subClassExpression super-
ClassExpression ')'
subClassExpression := ClassExpression
superClassExpression := ClassExpression

// *MOWL Media based axioms* //
MowlMediaAxiom := ConceptAxiom | MediaPropertyAxiom | PropagateAx-
iom

// *Property to associate media manifestations to MOWL concepts* //
ConceptAxiom := 'hasConcept' '(' axiomAnnotations BasicClassExpression
Concept ')'

// *Properties that allow association of media properties to concepts* //
MediaPropertyAxiom := MediaPatternAxiom | MediaExampleAxiom
MediaPatternAxiom := 'hasMediaPattern' '(' axiomAnnotations Concept Me-
diaPattern ')'
MediaExampleAxiom := 'hasMediaExample' '(' axiomAnnotations Concept
MediaExample ')'

// *Property that allows propagation of media properties between concepts* //
PropagateAxiom := 'propagateMedia' '(' axiomAnnotations Concept Concept ')'

// *Data Properties that link media patterns or media examples to URIs* //
MediaURIAxiom := hasURI' '(' axiomAnnotations MediaClassExpression xsd:anyURI ')'

// *Spatio-temporal properties allow special encoding for spatial, temporal and spatio-temporal associations* //
MowlSTAxiom := STAxiom | AllenAxiom | ContainmentAxiom | FuzzyAxiom
STAxiom := PredicateAxiom | ComponentAxiom
PredicateAxiom := 'has_predicate' '(' axiomAnnotations ComplexConcept Predicate ')'
ComponentAxiom := ComponentPropertyExpression '(' axiomAnnotations ComplexConcept Concept ')'

// *Fuzzy aspects of Spatio-temporal properties* //
AllenAxiom := AllenPropertyExpression '(' axiomAnnotations Predicate FuzzyAllen ')'
ContainAxiom := ContainPropertyExpression '(' axiomAnnotations Predicate FuzzyContainment ')'
FuzzyAxiom := FuzzAllenAxiom | FuzzyContainAxiom
FuzzyAllenAxiom := AllenDataExpression '(' axiomAnnotations FuzzyAllen FuzzyVector ')'
FuzzyContainAxiom := ContainDataExpression '(axiomAnnotations FuzzyContain FuzzyVector ')'

// *Uncertainty Specification related Axioms* //
CPTAxiom := CPTableAxiom | CPTRowAxiom | ConditionAxiom | CPTDataAxiom
CPTableAxiom := 'hasCPT' '(' axiomAnnotations BasicClassExpression CPTable ')'
CPTRowAxiom := 'hasRow' '(' axiomAnnotations CPTable CPRow ')'
ConditionAxiom := 'conditionedOn' '(' axiomAnnotations CPTable BasicClassExpression')'
CPTDataAxiom := ParentAxiom | ValueAxiom
ParentAxiom := 'parentStates' '(' axiomAnnotations CPRow xsd:string ')'
ValueAxiom := 'probValues' '(' axiomAnnotations CPRow mowl:probabilityVector ')'
...
// *Generic OWL2 Assertions apply to MOWL per se* //

Appendix C

Examples of WSDL Specification for Media Pattern Detectors

The following WSDL specification defines a face-recognition module that can be referred to as a media detector in a MOWL ontology. Similar specifications can be written for other media detectors. The <WSDL:types> element of the schema can be used to embed MPEG-7 descriptors in the specification if required.

```xml
<?xml version="1.0"?>
<definitions name="faceRecognizer"
    targetNamespace="file:///ICD/detectors/faceRecognizer.wsdl"
    xmlns:tns="file:///ICD/detectors/faceRecognizer.wsdl"
    xmlns:xsd1="file:///ICD/detectors/faceRecognizer.xsd"
    xmlns:soap="http://schemas.xmlsoap.org/wsdl/soap/"
    xmlns="http://schemas.xmlsoap.org/wsdl/">

    <types>
      <schema targetNamespace="file:///ICD/detectors/
        faceRecognizer.xsd"
            xmlns="http://www.w3.org/2000/10/XMLSchema">
          <element name="faceRecognizeRequest">
            <complexType>
              <all>
                  <element name="faceImage" type="anyURI"/>
                  <element name="expectedFace" type="string"/>
              </all>
            </complexType>
          </element>
          <element name="beliefValue">
            <complexType>
              <all>
                  <element name="belief" type=
                        "&mowlDatatypes;0_to_1_type"/>
              </all>
            </complexType>
          </element>
      </schema>
```

```
    </types>

    <message name="RecognizeFaceInput">
        <part name="body"
                element="xsd1:postureRecognizeRequest"/>
    </message>
    <message name="RecognizeFaceOutput">
        <part name="body" element="xsd1:beliefValue"/>
    </message>

    <portType name="FaceRecognizerPortType">
        <operation name="RecognizeFace">
            <input message="tns:RecognizeFaceInput"/>
            <output message="tns:RecognizeFaceOutput"/>
        </operation>
    </portType>

    <binding name="FaceRecognizerSoapBinding" type=
                "tns:FaceRecognizerPortType">
        <soap:binding style="document" transport=
                "http://schemas.xmlsoap.org/soap/http"/>
        <operation name="RecognizeFace">
            <soap:operation soapAction=
                "file:///ICD/detectors/FaceRecognizer"/>
            <input>
                <soap:body use="literal"/>
            </input>
            <output>
                <soap:body use="literal"/>
            </output>
        </operation>
    </binding>

    <service name="FaceRecognizerService">
        <documentation>
                Posture Recognition Service
        </documentation>
        <port name="FaceRecognizerPort" binding=
                "tns:FaceRecognizerSoapBinding">
          <soap:address location=
                "file:///ICD/detectors/FaceRecognizer"/>
        </port>
    </service>
</definitions>
```

Bibliography

[1] Stuti Ajmani, Hiranmay Ghosh, Anupama Mallik, and Santanu Chaudhury. An ontology based personalized garment recommendation system. In *Proceedings of the 2013 IEEE/WIC/ACM International Joint Conferences on Web Intelligence (WI) and Intelligent Agent Technologies (IAT) - Volume 03*, WI-IAT '13, pages 17–20, Washington, DC, USA, 2013. IEEE Computer Society.

[2] Sherif Akoush and Ahmed Sameh. Mobile user movement prediction using bayesian learning for neural networks. In *Proceedings of the 2007 International Conference on Wireless Communications and Mobile Computing*, IWCMC '07, pages 191–196, New York, NY, USA, 2007. ACM.

[3] Daniel G. Aliaga, Elisa Bertino, and Stefano Valtolina. Decho—a framework for the digital exploration of cultural heritage objects. *J. Comput. Cult. Herit.*, 3(3):12:1–12:26, February 2011.

[4] James F. Allen. Maintaining knowledge about temporal intervals. *Commun. ACM*, 26(11):832–843, November 1983.

[5] James F. Allen. Expert systems. chapter Maintaining Knowledge About Temporal Intervals, pages 248–259. IEEE Computer Society Press, Los Alamitos, CA, USA, 1990.

[6] Apache Mahout. http://mahout.apache.org/. [Online; accessed 29 January 2014].

[7] Daniel Appelquist, Dan Brickley, Melvin Carvahlo, Renato Iannella, Alexandre Passant, Christine Perey, and Henry Story. A standards-based, open and privacy-aware social web. http://www.w3.org/2005/Incubator/socialweb/XGR-socialweb-20101206/, 2010.

[8] Franz Baader and Werner Nutt. Basic description logics. In Franz Baader, Diego Calvanese, Deborah L. McGuinness, Daniele Nardi, and Peter F. Patel-Schneider, editors, *The Description Logic Handbook*, pages 43–95. Cambridge University Press, New York, NY, USA, 2003.

[9] Liang Bai, Songyang Lao, Gareth J. F. Jones, and Alan F. Smeaton. Video semantic content analysis based on ontology. In *Proceedings of the*

International Machine Vision and Image Processing Conference, IMVIP
'07, pages 117–124, Washington, DC, USA, 2007. IEEE Computer Society.

[10] Romil Bansal, Radhika Kumaran, Diwakar Mahajan, Arpit Khurdiya,
Lipika Dey, and Hiranmay Ghosh. Twipix: a web magazine curated from
social media. In *Proceedings of the 20th ACM International Conference
on Multimedia*, MM '12, pages 1355–1356, New York, NY, USA, 2012.
ACM.

[11] Bing-Kun Bao, Weiqing Min, Jitao Sang, and Changsheng Xu. Multimedia news digger on emerging topics from social streams. In *Proceedings of
the 20th ACM International Conference on Multimedia*, MM '12, pages
1357–1358, New York, NY, USA, 2012. ACM.

[12] L. Barrington, A. Chan, D. Turnbull, and G. Lanckriet. Audio information retrieval using semantic similarity. In *IEEE International Conference on Acoustics, Speech and Signal Processing, 2007. ICASSP 2007.*,
volume 2, pages II–725–II–728, April 2007.

[13] David   Beckett,   Tim   Berners-Lee,   Eric   Prud'hommeaux,   and
Gavin   Carothers.   RDF   1.1   Turtle: Terse   RDF   triple   language.
http://www.w3.org/TR/turtle/, 2014.

[14] Perakath C. Benjamin, Christopher P. Menzel, Richard J. Mayer, Flo-
rence Fillion, Michael T. Futrell, Paula S. deWitte, and Madhavi Lingi-
neni. IDEF5 method report. http://www.idef.com/pdf/Idef5.pdf, 1994.

[15] Tim Berners-Lee, James Hendler, and Ora Lassila. The semantic web.
Scientific American, 284(5):35–43, May 2001.

[16] Stefano Berretti, Alberto Del Bimbo, and Pietro Pala. Merging results
for distributed content based image retrieval. *Multimedia Tools Appl.*,
24(3):215–232, December 2004.

[17] Marco Bertini, Alberto Del Bimbo, Giuseppe Serra, Carlo Torniai, Rita
Cucchiara, Costantino Grana, and Roberto Vezzani. Dynamic pictori-
ally enriched ontologies for digital video libraries. *IEEE MultiMedia*,
16(2):42–51, April 2009.

[18] Marco Bertini, Alberto Del Bimbo, and Giuseppe Serra. Learning ontol-
ogy rules for semantic video annotation. In *Proceedings of the 2nd ACM
Workshop on Multimedia Semantics*, MS '08, pages 1–8, New York, NY,
USA, 2008. ACM.

[19] David M. Blei, Andrew Y. Ng, and Michael I. Jordan. Latent dirichlet
allocation. *J. Mach. Learn. Res.*, 3:993–1022, March 2003.

[20] Avrim Blum and Tom Mitchell. Combining labeled and unlabeled data with co-training. In *Proceedings of the Eleventh Annual Conference on Computational Learning Theory*, COLT' 98, pages 92–100, New York, NY, USA, 1998. ACM.

[21] Conrad Bock, Achille Fokoue, Peter Haase, Rinke Hoekstra, Ian Horrocks, Alan Ruttenberg, Uli Sattler, and Michael Smith. OWL 2 web ontology language: Structural specification and functional-style syntax (second edition). http://www.w3.org/TR/owl2-syntax/, 2012.

[22] Anna Bosch, Andrew Zisserman, and Xavier Muñoz. Scene classification using a hybrid generative/discriminative approach. *IEEE Trans. Pattern Anal. Mach. Intell.*, 30(4):712–727, April 2008.

[23] Leo Breiman. Random forests. *Mach. Learn.*, 45(1):5–32, October 2001.

[24] Alexander Budanitsky and Alexander Budanitsky. Lexical semantic relatedness and its application in natural language processing. ftp://ftp.cs.toronto.edu/pub/gh/Budanitsky-99.pdf, 1999.

[25] Paul Buitelaar and Bernardo Magnini. Ontology learning from text: An overview. In Paul Buitelaar, Cimiano P., and Magnini B. P., editors, *Ontology Learning from Text: Methods, Applications and Evaluation*, pages 3–12. IOS Press, 2005.

[26] Horst Bunke. Graph matching: Theoretical foundations, algorithms, and applications. In *International Conference on Vision Interface*, pages 82–84, May 2000.

[27] Walter Carrer-Neto, María Luisa Hernández-Alcaraz, Rafael Valencia-García, and Francisco García-Sánchez. Social knowledge-based recommender system. application to the movies domain. *Expert Syst. Appl.*, 39(12):10990–11000, September 2012.

[28] Lois Mai Chan. *Cataloging and Classification: An Introduction*. McGraw-Hill, second edition, 1994.

[29] Shih-Fu Chang, Thomas Sikora, and Atul Purl. Overview of the MPEG-7 standard. *IEEE Transactions on Circuits and Systems for Video Technology*, 11(6):688–695, 2001.

[30] Yung-Chun Chang, Pei-Ching Yang, and Jung-Hsien Chiang. Ontology-based intelligent web mining agent for taiwan travel. In *Proceedings of the 2009 IEEE/WIC/ACM International Joint Conference on Web Intelligence and Intelligent Agent Technology - Volume 03*, WI-IAT '09, pages 421–424, Washington, DC, USA, 2009. IEEE Computer Society.

[31] Kasturi Chatterjee, S. Masoud Sadjadi, and Shu-Ching Chen. *A Distributed Multimedia Data Management over the Grid*, pages 27–48. Springer Berlin Heidelberg, 2010.

[32] Liming Chen and Chris Nugent. Ontology-based activity recognition in intelligent pervasive environments. *Int. J. Web Inf. Sys.*, 5(4):410–430, 2009.

[33] Ming-yu Chen, Michael Christel, Alexander Hauptmann, and Howard Wactlar. Putting active learning into multimedia applications: Dynamic definition and refinement of concept classifiers. In *Proceedings of the 13th Annual ACM International Conference on Multimedia*, MULTIMEDIA '05, pages 902–911, New York, NY, USA, 2005. ACM.

[34] Wei-Ta Chu and Ming-Hung Tsai. Visual pattern discovery for architecture image classification and product image search. In *Proceedings of the 2nd ACM International Conference on Multimedia Retrieval*, ICMR '12, pages 27:1–27:8, New York, NY, USA, 2012. ACM.

[35] L. Cinque, G. Ciocca, S. Levialdi, A. Pellicano, and R. Schettini. Color-based image retrieval using spatial-chromatic histograms. *Image and Vision Computing*, 19(13):979–986, 2001.

[36] Michael Compton, Payam Barnaghi, Luis Bermudez, Raúl García-Castro, Oscar Corcho, Simon Cox, John Graybeal, Manfred Hauswirth, Cory Henson, Arthur Herzog, Vincent Huang, Krzysztof Janowicz, W. David Kelsey, Danh Le Phuoc, Laurent Lefort, Myriam Leggieri, Holger Neuhaus, Andriy Nikolov, Kevin Page, Alexandre Passant, Amit Sheth, and Kerry Taylor. Ontology paper: The SSN ontology of the W3C semantic sensor network incubator group. *Web Semant.*, 17:25–32, December 2012.

[37] Owen Conlan, Ian O'Keeffe, and Shane Tallon. Combining adaptive hypermedia techniques and ontology reasoning to produce dynamic personalized news services. In *Proceedings of the 4th International Conference on Adaptive Hypermedia and Adaptive Web-Based Systems*, AH'06, pages 81–90, Berlin, Heidelberg, 2006. Springer-Verlag.

[38] Arturo Crespo and Hector Garcia-Molina. Semantic overlay networks for P2P systems. In *Proceedings of the Third International Conference on Agents and Peer-to-Peer Computing*, AP2PC'04, pages 1–13, Berlin, Heidelberg, 2005. Springer-Verlag.

[39] Isabel F. Cruz and Kimberly M. James. A user interface for distributed multimedia database querying with mediator supported refinement. In *Proceedings of the 1999 International Symposium on Database Engineering & Applications*, IDEAS '99, pages 433–, Washington, DC, USA, 1999. IEEE Computer Society.

[40] Gabriella Csurka, Christopher R. Dance, Lixin Fan, Jutta Willamowski, and Cdric Bray. Visual categorization with bags of keypoints. In *In Workshop on Statistical Learning in Computer Vision, ECCV*, pages 1–22, 2004.

[41] Navneet Dalal and Bill Triggs. Histograms of oriented gradients for human detection. In *IEEE Computer Society Conference on Computer Vision and Pattern Recognition, 2005. CVPR 2005.*, volume 1, pages 886–893. IEEE, 2005.

[42] Kareem Darwish and Walid Magdy. Error correction vs. query garbling for Arabic OCR document retrieval. *ACM Trans. Inf. Syst.*, 26(1), November 2007.

[43] Stamatia Dasiopoulou, Vassilis Tzouvaras, Ioannis Kompatsiaris, and Michael G. Strintzis. Enquiring MPEG-7 based multimedia ontologies. *Multimedia Tools Appl.*, 46(2-3):331–370, January 2010.

[44] J. Dass, M. Sharma, E. Hassan, and H. Ghosh. A density based method for automatic hairstyle discovery and recognition. In *Computer Vision, Pattern Recognition, Image Processing and Graphics (NCVPRIPG), 2013 Fourth National Conference on*, pages 1–4, Dec 2013.

[45] Alain De Cheveigné and Hideki Kawahara. YIN, a fundamental frequency estimator for speech and music. *J. Acous. Soc. Am.*, 111(4):1917–1930, 2002.

[46] Jeffrey Dean and Sanjay Ghemawat. MapReduce: Simplified data processing on large clusters. *Commun. ACM*, 51(1):107–113, January 2008.

[47] Jia Deng, Nan Ding, Yangqing Jia, Andrea Frome, Kevin Murphy, Samy Bengio, Yuan Li, Hartmut Neven, and Hartwig Adam. Large-scale object classification using label relation graphs. In *Computer Vision–ECCV 2014*, pages 48–64. Springer, 2014.

[48] Zhongli Ding and Yun Peng. A probabilistic extension to ontology language owl. In *Proceedings of the Proceedings of the 37th Annual Hawaii International Conference on System Sciences (HICSS'04) - Track 4 - Volume 4*, HICSS '04, pages 40111.1–, Washington, DC, USA, 2004. IEEE Computer Society.

[49] Martin Doerr. The CIDOC conceptual reference module: An ontological approach to semantic interoperability of metadata. *AI Mag.*, 24(3):75–92, September 2003.

[50] S. T. Dumais, G. W. Furnas, T. K. Landauer, S. Deerwester, and R. Harshman. Using latent semantic analysis to improve access to textual information. In *Proceedings of the SIGCHI Conference on Human Factors in Computing Systems*, CHI '88, pages 281–285, New York, NY, USA, 1988. ACM.

[51] Marc Ehrig. *Ontology Alignment: Bridging the Semantic Gap (Semantic Web and Beyond)*. Springer-Verlag New York, Inc., Secaucus, NJ, USA, 2006.

[52] Marc Ehrig. *Ontology Alignment: Bridging the Semantic Gap*, volume 4 of *Semantic Web And Beyond Computing for Human Experience*. Springer, 2007.

[53] Slim Essid, Gaël Richard, and Bertrand David. Instrument recognition in polyphonic music based on automatic taxonomies. *IEEE Trans. Audio, Speech, and Language Processing*, 14(1):68–80, 2006.

[54] Emmanuel Eze, Tanko Ishaya, and Dawn Wood. Contextualizing multimedia semantics towards personalised elearning. *J. Digital Information Management*, 5(2):61, 2007.

[55] Li Fei-Fei and Pietro Perona. A bayesian hierarchical model for learning natural scene categories. In *IEEE Computer Society Conference on Computer Vision and Pattern Recognition, 2005. CVPR 2005*, volume 2, pages 524–531. IEEE, 2005.

[56] Pedro Felzenszwalb and Daniel Huttenlocher. Distance transforms of sampled functions. Technical report, Cornell University, 2004.

[57] Pedro F Felzenszwalb, Ross B Girshick, David McAllester, and Deva Ramanan. Object detection with discriminatively trained part-based models. *IEEE Trans. Pattern Analysis and Machine Intelligence*, 32(9):1627–1645, 2010.

[58] Steve Feng, Romain Caire, Bingen Cortazar, Mehmet Turan, Andrew Wong, and Aydogan Ozcan. Immunochromatographic diagnostic test analysis using Google Glass. *ACS Nano*, 8(3):3069–3079, 2014.

[59] Myron Flickner, Harpreet Sawhney, Wayne Niblack, Jonathan Ashley, Qian Huang, Byron Dom, Monika Gorkani, Jim Hafner, Denis Lee, Dragutin Petkovic, et al. Query by image and video content: The QBIC system. *Computer*, 28(9):23–32, 1995.

[60] James D. Foley, Andries van Dam, Steven K. Feiner, and John F. Hughes. *Computer Graphics: Principles and Practice*. Addison-Wesley Professional, 1982.

[61] Nir Friedman and Moises Goldszmidt. Learning bayesian networks with local structure. In *Proceedings of the Twelfth International Conference on Uncertainty in Artificial Intelligence*, UAI'96, pages 252–262, San Francisco, CA, USA, 1996. Morgan Kaufmann Publishers Inc.

[62] Nir Friedman and Zohar Yakhini. On the sample complexity of learning bayesian networks. In *Proceedings of the Twelfth International Conference on Uncertainty in Artificial Intelligence*, UAI'96, pages 274–282, San Francisco, CA, USA, 1996. Morgan Kaufmann Publishers Inc.

[63] Tianshi Gao and Daphne Koller. Discriminative learning of relaxed hierarchy for large-scale visual recognition. In *Proceedings of the 2011 International Conference on Computer Vision*, ICCV '11, pages 2072–2079, Washington, DC, USA, 2011. IEEE Computer Society.

[64] Nikhil Garg and Daniel Gatica-Perez. Tagging and retrieving images with co-occurrence models: From corel to flickr. In *Proceedings of the First ACM Workshop on Large-Scale Multimedia Retrieval and Mining*, LS-MMRM '09, pages 105–112, New York, NY, USA, 2009. ACM.

[65] Hiranmay Ghosh. *R-MAGIC: A cooperative agent based architecture for retrieval of multimedia documents distributed over heterogeneous repositories*. PhD thesis, Indian Institute of Technology, Delhi, 2003.

[66] Hiranmay Ghosh and Santanu Chaudhury. Guided shopping in e-market. In *Proceedings of Knowledge Based Computer Systems (KBCS)*, 2002.

[67] Hiranmay Ghosh and Santanu Chaudhury. Distributed and reactive query planning in R-MAGIC: An agent-based multimedia retrieval system. *IEEE Trans. Knowl. Data Eng.*, 16(9):1082–1095, September 2004.

[68] Hiranmay Ghosh, Sunil Kumar Kopparapu, Tanushyam Chattopadhyay, Ashish Khare, Sujal Subhash Wattamwar, Amarendra Gorai, and Meghna Pandharipande. Multimodal indexing of multilingual news video. *Int. J. Digital Multimedia Broadcasting*, 2010.

[69] Ross Girshick, Jeff Donahue, Trevor Darrell, and Jitendra Malik. Rich feature hierarchies for accurate object detection and semantic segmentation. *arXiv preprint arXiv:1311.2524*, 2013.

[70] Abhinav Goel, Mayank Juneja, and C. V. Jawahar. Are buildings only instances?: Exploration in architectural style categories. In *Proceedings of the Eighth Indian Conference on Computer Vision, Graphics and Image Processing*, ICVGIP '12, pages 1:1–1:8, New York, NY, USA, 2012. ACM.

[71] Ben Gold, Nelson Morgan, and Dan Ellis. *Speech and Audio Signal Processing: Processing and Perception of Speech and Music*. John Wiley & Sons, 2011.

[72] Kristen Grauman and Trevor Darrell. The pyramid match kernel: Discriminative classification with sets of image features. In *Proceedings of the Tenth IEEE International Conference on Computer Vision - Volume 2*, ICCV '05, pages 1458–1465, Washington, DC, USA, 2005. IEEE Computer Society.

[73] Alasdair J. G. Gray, Raúl García-Castro, Kostis Kyzirakos, Manos Karpathiotakis, Jean-Paul Calbimonte, Kevin Page, Jason Sadler, Alex

Frazer, Ixent Galpin, Alvaro A. A. Fernandes, Norman W. Paton, Oscar Corcho, Manolis Koubarakis, David De Roure, Kirk Martinez, and Asunción Gómez-Pérez. A semantically enabled service architecture for mashups over streaming and stored data. In *Proceedings of the 8th Extended Semantic Web Conference on The Semanic Web: Research and Applications - Volume Part II*, ESWC'11, pages 300–314, Berlin, Heidelberg, 2011. Springer-Verlag.

[74] Alasdair J. G. Gray, Jason Sadler, Oles Kit, Kostis Kyzirakos, Manos Karpathiotakis, Jean-Paul Calbimonte, Kevin Page, Raúl García-Castro, Alex Frazer, Ixent Galpin, Alvaro A. A. Fernandes, Norman W. Paton, Oscar Corcho, Manolis Koubarakis, David De Roure, Kirk Martinez, and Asunción Gómez-Pérez. A semantic sensor web for environmental decision support applications. *Sensors*, 11(9):8855–8887, 2011.

[75] William I. Grosky, D. V. Sreenath, Farshad Fotouhi, and Vlad Wietrzyk. The multimedia semantic web. *Int. J. Comput. Sci. Eng.*, 2(5/6):326–340, August 2006.

[76] CCTV User Group. How many cameras are there? https://www.cctvusergroup.com/art.php?art=94, (last accessed: September 2014).

[77] Network Working Group. Internationalized Resource Identifier. http://tools.ietf.org/html/rfc3987.

[78] Network Working Group. Universal Resource Identifier. http://tools.ietf.org/html/rfc3986.

[79] Thomas R. Gruber. Toward principles for the design of ontologies used for knowledge sharing. *Int. J. Hum.-Comput. Stud.*, 43(5–6):907–928, December 1995.

[80] Guodong Guo and S. Z. Li. Content-based audio classification and retrieval by support vector machines. *Trans. Neur. Netw.*, 14(1):209–215, January 2003.

[81] Samira Hammiche, Salima Benbernou, Mohand-Saïd Hacid, and Athena Vakali. Semantic retrieval of multimedia data. In *Proceedings of the 2nd ACM International Workshop on Multimedia Databases*, MMDB '04, pages 36–44, New York, NY, USA, 2004. ACM.

[82] R.M. Haralick, K. Shanmugam, and Its'Hak Dinstein. Textural features for image classification. *IEEE Trans. Systems, Man and Cybernetics*, SMC-3(6):610–621, Nov 1973.

[83] Gaurav Harit. *On Harnessing Video Content*. PhD thesis, Indian Institute of Technology, Delhi, 2007.

[84] Gaurav Harit, Santanu Chaudhury, and Hiranmay Ghosh. Managing document images in a digital library: An ontology guided approach. In *Proceedings of the First International Workshop on Document Image Analysis for Libraries (DIAL'04)*, DIAL '04, pages 64–, Washington, DC, USA, 2004. IEEE Computer Society.

[85] Ehtesham Hassan, Santanu Chaudhury, and M. Gopal. Document image indexing using edit distance based hashing. In *Proceedings of the 2011 International Conference on Document Analysis and Recognition*, ICDAR '11, pages 1200–1204, Washington, DC, USA, 2011. IEEE Computer Society.

[86] Ehtesham Hassan, Santanu Chaudhury, and M. Gopal. Multi-modal information integration for document retrieval. In *Proceedings of the 2013 12th International Conference on Document Analysis and Recognition*, ICDAR '13, pages 1200–1204, Washington, DC, USA, 2013. IEEE Computer Society.

[87] Ehtesham Hassan, Hiranmay Ghosh, Rajendra Nagar, and Stuti Ajmani. An ontology based personalized recommendation system for ethnic garments. *Human-centric Computing and Information Sciences*, In Press, (accepted 2014).

[88] Maryam Hazman, Samhaa R. El-Beltagy, and Ahmed Rafea. A survey of ontology learning approaches. *Int. J. of Comput. App.*, 22(8):36–43, May 2011.

[89] Marti A. Hearst. Untangling text data mining. In *Proceedings of the 37th Annual Meeting of the Association for Computational Linguistics on Computational Linguistics*, ACL '99, pages 3–10, Stroudsburg, PA, USA, 1999. Association for Computational Linguistics.

[90] David Heckerman. Learning in graphical models. chapter A Tutorial on Learning with Bayesian Networks, pages 301–354. MIT Press, Cambridge, MA, USA, 1999.

[91] Francisca Hernendez. Semantic web use cases and case studies: An ontology of Cantabria's cultural heritage. http://www.w3.org/2001/sw/sweo/public/UseCases/FoundationBotin/, 2007.

[92] Jerry Hobbs. Influences and inferences. *Comput. Linguist.*, 39(4):781–798, December 2013.

[93] Thomas Hofmann. Probabilistic latent semantic indexing. In *Proceedings of the 22nd Annual International ACM SIGIR Conference on Research and Development in Information Retrieval*, SIGIR '99, pages 50–57, New York, NY, USA, 1999. ACM.

[94] Ian Horrocks, Peter F. Patel-Schneider, and Frank van Harmelen. From SHIQ and RDF to OWL: The making of a web ontology language. *J. Web Sem.*, 1(1):7–26, 2003.

[95] Xuedong Huang, Alex Acero, and Hsiao-Wuen Hon. *Spoken Language Processing: A Guide to Theory, Algorithm, and System Development*. Prentice Hall PTR, Upper Saddle River, NJ, USA, 1st edition, 2001.

[96] David H. Hubel and Torsten N. Wiesel. *Brain and Visual Perception: The Story of a 25-Year Collaboration*. Oxford University Press, 2004.

[97] Michael N. Huhns and Munindar P. Singh. Ontologies for agents. *IEEE Internet Computing*, 1(6):81–83, November 1997.

[98] Eric J. Humphrey, Juan P. Bello, and Yann Lecun. Feature learning and deep architectures: New directions for music informatics. *J. Intell. Inf. Syst.*, 41(3):461–481, December 2013.

[99] Jane Hunter. Combining the CIDOC CRM and MPEG-7 to describe multimedia in museums. In *Museums and the Web 2002*. Museum Publications, 2002.

[100] Jane Hunter. Enhancing the semantic interoperability of multimedia through a core ontology. *IEEE Trans. Cir. and Sys. for Video Technol.*, 13(1):49–58, January 2003.

[101] Jane Hunter. *Adding Multimedia to the Semantic Web-Building and Applying an MPEG-7 Ontology*. Wiley, 2005.

[102] Katsushi Ikeuchi, Takeshi Oishi, Masataka Kagesawa, Atsuhiko Banno, Rei Kawakami, Tetsuya Kakuta, Yasuhide Okamoto, and Boun Vinh Lu. Outdoor gallery and its photometric issues. In *Proceedings of the 9th ACM SIGGRAPH Conference on Virtual Reality Continuum and its Applications in Industry*, VRCAI '10, pages 361–364, New York, NY, USA, 2010. ACM.

[103] Daniel Isemann and Khurshid Ahmad. Ontological access to images of fine art. *J. Comput. Cult. Herit.*, 7(1):3:1–3:25, April 2014.

[104] Tanko Ishaya and Emmanuel Eze. Context-based multimedia ontology model. In *Proceedings of the Third International Conference on Autonomic and Autonomous Systems*, ICAS '07, pages 2–, Washington, DC, USA, 2007. IEEE Computer Society.

[105] A. Jaimes and J. R. Smith. Semi-automatic, data-driven construction of multimedia ontologies. In *Proceedings of the 2003 International Conference on Multimedia and Expo - Volume 2*, ICME '03, pages 781–784, Washington, DC, USA, 2003. IEEE Computer Society.

[106] C. V. Jawahar, A. Balasubramanian, Million Meshesha, and Anoop M. Namboodiri. Retrieval of online handwriting by synthesis and matching. *Pattern Recogn.*, 42(7):1445–1457, July 2009.

[107] Jayadeva, R. Khemchandani, and S. Chandra. Twin support vector machines for pattern classification. *IEEE Trans. Pattern Analysis and Machine Intelligence*, 29(5):905–910, May 2007.

[108] E. T. Jaynes. *Probability Theory: The Logic of Science.* Cambridge University Press, 2003.

[109] Anubha Jindal, Aditya Tiwari, and Hiranmay Ghosh. Efficient and language independent news story segmentation for telecast news videos. In *Proceedings of the 2011 IEEE International Symposium on Multimedia*, ISM '11, pages 458–463, Washington, DC, USA, 2011. IEEE Computer Society.

[110] Dan Jurafsky and James H Martin. *Speech and Language Processing: An Introduction to Natural Language Processing, Computational Linguistics, and Speech Recognition.* Prentice Hall, second edition, 2008.

[111] B. Kentner. *Color Me a Season: A Complete Guide to Finding Your Best Colors and How to Use Them.* Ken Kra Publishers, 1979.

[112] Arpit Khurdiya, Lipika Dey, Diwakar Mahajan, and Ishan Verma. Extraction and compilation of events and sub-events from twitter. In *Proceedings of the The 2012 IEEE/WIC/ACM International Joint Conferences on Web Intelligence and Intelligent Agent Technology - Volume 01*, WI-IAT '12, pages 504–508, Washington, DC, USA, 2012. IEEE Computer Society.

[113] Christoph Carl Kling, Jérôme Kunegis, Sergej Sizov, and Steffen Staab. Detecting non-gaussian geographical topics in tagged photo collections. In *Proceedings of the 7th ACM International Conference on Web Search and Data Mining*, WSDM '14, pages 603–612, New York, NY, USA, 2014. ACM.

[114] Matthias Klusch, Stefano Lodi, and Gianluca Moro. Agent-based distributed data mining: The KDEC scheme. In Matthias Klusch, Sonia Bergamaschi, Pete Edwards, and Paolo Petta, editors, *Intelligent Information Agents*, pages 104–122. Springer-Verlag, Berlin, Heidelberg, 2003.

[115] Daphne Koller and Nir Friedman. *Probabilistic Graphical Models.* MIT Press, 2009.

[116] Konstantinos Kotis and Andreas Papasalouros. Automated learning of social ontologies. In Wilson Wong, Wei Liu, and Mohammed Bennamoun, editors, *Ontology Learning and Knowledge Discovery Using*

the Web: Challenges and Recent Advances, chapter 12, pages 227–246. IGI-Global, 2011.

[117] Wessel Kraaij, Alan F Smeaton, and Paul Over. Trecvid 2004 – an overview. In *Proceedings of TRECVID 2004*. NIST, USA, 2004.

[118] Alex Krizhevsky, Ilya Sutskever, and Geoffrey E Hinton. ImageNet classification with deep convolutional neural networks. In *Advances in Neural Information Processing Systems*, pages 1097–1105, 2012.

[119] Honglak Lee, Roger Grosse, Rajesh Ranganath, and Andrew Y. Ng. Unsupervised learning of hierarchical representations with convolutional deep belief networks. *Commun. ACM*, 54(10):95–103, October 2011.

[120] Jens Lehmann and Johanna Voelker. An introduction to ontology learning. In Jens Lehmann and Johanna Voelker, editors, *Perspectives on Ontology Learning*, pages ix–xvi. AKA/IOS Press, 2014.

[121] Cane Wing-ki Leung, Stephen Chi-fai Chan, and Fu-lai Chung. An empirical study of a cross-level association rule mining approach to cold-start recommendations. *Knowl.-Based Syst.*, 21(7):515–529, October 2008.

[122] Yuan-Fang Li, Gavin Kennedy, Faith Davies, and Jane Hunter. PODD: An ontology-driven data repository for collaborative phenomics research. In *Proceedings of the Role of Digital Libraries in a Time of Global Change, and 12th International Conference on Asia-Pacific Digital Libraries*, ICADL'10, pages 179–188, Berlin, Heidelberg, 2010. Springer-Verlag.

[123] Si Liu, Jiashi Feng, Zheng Song, Tianzhu Zhang, Hanqing Lu, Changsheng Xu, and Shuicheng Yan. Hi, magic closet, tell me what to wear! In *Proceedings of the 20th ACM International Conference on Multimedia*, MM '12, pages 619–628, New York, NY, USA, 2012. ACM.

[124] David G. Lowe. Distinctive image features from scale-invariant keypoints. *Int. J. Comput. Vision*, 60(2):91–110, November 2004.

[125] Lie Lu, Hong-Jiang Zhang, and Hao Jiang. Content analysis for audio classification and segmentation. *IEEE Trans. on Speech and Audio Processing*, 10(7):504–516, Oct 2002.

[126] Alexander Maedche and Steffen Staab. Ontology learning for the semantic web. *IEEE Intell. Syst.*, 16(2):72–79, March 2001.

[127] Bridaduti Maiti, Hiranmay Ghosh, and Santanu Chaudhury. A framework for ontology specification and integration for multimedia applications. In *Proceedings of Knowledge Based Computer System (KBCS 2004)*, December 2004.

[128] Anupama Mallik, Santanu Chaudhury, T. B. Dinesh, and Chaluvaraju. An intellectual journey in history: Preserving indian cultural heritage. In *New Trends in Image Analysis and Processing - ICIAP 2013 - ICIAP 2013 International Workshops, Naples, Italy, September 9-13, 2013. Proceedings*, pages 298–307, 2013.

[129] Anupama Mallik, Santanu Chaudhury, and Hiranmay Ghosh. Nrityakosha: Preserving the intangible heritage of indian classical dance. *J. Comput. Cult. Herit.*, 4(3):11:1–11:25, December 2011.

[130] Anupama Mallik, Santanu Chaudhury, Shipra Madan, T. B. Dinesh, and Uma V. Chandru. Archiving mural paintings using an ontology based approach. In *ACCV 2012 International Workshops, Daejeon, Korea*, volume 7729, pages 37–48. Springer Berlin Heidelberg, 2012.

[131] Anupama Mallik, Hiranmay Ghosh, Santanu Chaudhury, and Gaurav Harit. MOWL: An ontology representation language for web-based multimedia applications. *ACM Trans. Multimedia Comput. Commun. Appl.*, 10(1):8:1–8:21, December 2013.

[132] John Markel. The SIFT algorithm for fundamental frequency estimation. *IEEE Trans. Audio and Electroacoustics*, 20(5):367–377, 1972.

[133] Marcin Marszałek and Cordelia Schmid. Constructing category hierarchies for visual recognition. In *Proceedings of the 10th European Conference on Computer Vision: Part IV*, ECCV '08, pages 479–491, Berlin, Heidelberg, 2008. Springer-Verlag.

[134] Jose M. Martinez. MPEG-7: Overview of MPEG-7 description tools, part 2. *IEEE MultiMedia*, 9(3):83–93, July 2002.

[135] George A. Miller. WordNet: A lexical database for English. *Commun. ACM*, 38(11):39–41, November 1995.

[136] G. Mori, S. Belongie, and J. Malik. Shape contexts enable efficient retrieval of similar shapes. In *Proceedings of the 2001 IEEE Computer Society Conference on Computer Vision and Pattern Recognition, 2001. CVPR 2001.*, volume 1, pages I–723–I–730 vol.1, 2001.

[137] Phivos Mylonas, Thanos Athanasiadis, Manolis Wallace, Yannis Avrithis, and Stefanos Kollias. Semantic representation of multimedia content: Knowledge representation and semantic indexing. *Multimedia Tools Appl.*, 39(3):293–327, September 2008.

[138] M. Naphade, J. R. Smith, J. Tesic, Shih-Fu Chang, W. Hsu, L. Kennedy, A. Hauptmann, and J. Curtis. Large-scale concept ontology for multimedia. *MultiMedia, IEEE*, 13(3):86–91, July 2006.

[139] M.R. Naphade and J.R. Smith. Active learning for simultaneous annotation of multiple binary semantic concepts [video content analysis]. In *Multimedia and Expo, 2004. ICME '04. 2004 IEEE International Conference on*, volume 1, pages 77–80 Vol.1, June 2004.

[140] Roberto Navigli, Paola Velardi, and Aldo Gangemi. Ontology learning and its application to automated terminology translation. *IEEE Intelligent Systems*, 18(1):22–31, 2003.

[141] Richard E. Neapolitan. *Probabilistic Reasoning in Expert Systems: Theory and Alogorithms*. John Wiley and Sons, Inc, 1990.

[142] Radu Andrei Negoescu and Daniel Gatica-Perez. Analyzing flickr groups. In *Proceedings of the 2008 International Conference on Content-based Image and Video Retrieval*, CIVR '08, pages 417–426, New York, NY, USA, 2008. ACM.

[143] Radu Andrei Negoescu, Alexander C. Loui, and Daniel Gatica-Perez. Kodak moments and flickr diamonds: How users shape large-scale media. In *Proceedings of the International Conference on Multimedia*, MM '10, pages 1027–1030, New York, NY, USA, 2010. ACM.

[144] Lukasz Neuman, Jakub Kozlowski, and Aleksander Zgrzywa. Information retrieval using bayesian networks. *Computational Science - ICCS 2004*, pages 521–528, 2004.

[145] Radu Stefan Niculescu, Tom M. Mitchell, and R. Bharat Rao. Bayesian network learning with parameter constraints. *J. Mach. Learn. Res.*, 7:1357–1383, December 2006.

[146] Alexandros Ntousias, Nektarios Gioldasis, Chrisa Tsinaraki, and Stavros Christodoulakis. Rich metadata and context capturing through cidoc/crm and mpeg-7 interoperability. In *Proceedings of the 2008 International Conference on Content-based Image and Video Retrieval*, CIVR '08, pages 151–160, New York, NY, USA, 2008. ACM.

[147] R. Odon de Alencar, L. E. Zarate, and M. A. J. Song. Sphere-M: An ontology capture method. In *2012 IEEE International Conference on Systems, Man, and Cybernetics (SMC)*, pages 353–358, Oct 2012.

[148] Paul Over, Tzveta Ianeva, Wessel Kraaij, and Alan F Smeaton. Trecvid 2005 – an overview. In *Proceedings of TRECVID 2005*. NIST,USA, 2005.

[149] N. Pahal, S. Chaudhury, and B. Lall. Ontology driven contextual tagging of multimedia data. In *2014 IEEE International Conference on Multimedia and Expo Workshops (ICMEW)*, pages 1–6, July 2014.

[150] Nisha Pahal, Santanu Chaudhury, and Brejesh Lall. Extending MOWL for event representation (E-MOWL). In *Proceedings of the 2013 IEEE/WIC/ACM International Joint Conferences on Web Intelligence (WI) and Intelligent Agent Technologies (IAT) - Volume 03*, WI-IAT '13, pages 171–174, Washington, DC, USA, 2013. IEEE Computer Society.

[151] Georgios Paliouras, Constantine D. Spyropoulos, and George Tsatsaronis. Bootstrapping ontology evolution with multimedia information extraction. In Georgios Paliouras, Constantine D. Spyropoulos, and George Tsatsaronis, editors, *Knowledge-Driven Multimedia Information Extraction and Ontology Evolution*, pages 1–17. Springer-Verlag, Berlin, Heidelberg, 2011.

[152] Xueming Pan, Thomas Schiffer, Martin Schröttner, René Berndt, Martin Hecher, Sven Havemann, and Dieter W. Fellner. An enhanced distributed repository for working with 3D assets in cultural heritage. In *Proceedings of the 4th International Conference on Progress in Cultural Heritage Preservation*, EuroMed'12, pages 349–358, Berlin, Heidelberg, 2012. Springer-Verlag.

[153] Dimitris Papadias, Nikos Mamoulis, and Vasilis Delis. Approximate spatio-temporal retrieval. *ACM Trans. Inf. Syst.*, 19(1):53–96, January 2001.

[154] What is your color season? spring, summer, autumn, winter? http://personalitycafe.com/general-chat/37461-what-your-color-season-spring-summer-autumn-winter-fun.html (last accessed: September, 2014).

[155] K. Petridis, S. Bloehdorn, C. Saathoff, N. Simou, S. Dasiopoulou, V. Tzouvaras, S. Handschuh, Y. Avrithis, Y. Kompatsiaris, and S. Staab. Knowledge representation and semantic annotation for multimedia analysis and reasoning. *IEE Proceedings – Vision, Image and Signal Processing*, 153(3):255–262, 2006.

[156] J. Philbin, O. Chum, M. Isard, J. Sivic, and A. Zisserman. Object retrieval with large vocabularies and fast spatial matching. In *IEEE Conference on Computer Vision and Pattern Recognition, 2007. CVPR '07.*, pages 1–8, June 2007.

[157] Internet 2012 in numbers. http://royal.pingdom.com/2013/01/16/internet-2012-in-numbers/, 2013.

[158] R. Porter and N. Canagarajah. Robust rotation-invariant texture classification: wavelet, Gabor filter and GMRF based schemes. *IEE Proceedings – Vision, Image and Signal Processing*, 144(3):180–188, Jun 1997.

[159] Malcolm Pradhan, Max Henrion, Gregory Provan, Brendan Del Favero, and Kurt Huang. The sensitivity of belief networks to imprecise probabilities: An experimental investigation. *Artif. Intell.*, 85(1-2):363–397, August 1996.

[160] Mohamad Rabbath, Philipp Sandhaus, and Susanne Boll. Automatic creation of photo books from stories in social media. *ACM Trans. Multimedia Comput. Commun. Appl.*, 7S(1):27:1–27:18, November 2011.

[161] Lawrence Rabiner. A tutorial on hidden Markov models and selected applications in speech recognition. *Proceedings of the IEEE*, 77(2):257–286, 1989.

[162] Lawrence R Rabiner and Biing-Hwang Juang. *Fundamentals of Speech Recognition*, volume 14. PTR Prentice Hall Englewood Cliffs, 1993.

[163] Setareh Rafatirad, Amarnath Gupta, and Ramesh Jain. Event composition operators: ECO. In *Proceedings of the 1st ACM international workshop on Events in multimedia*, EiMM '09, pages 65–72, New York, NY, USA, 2009. ACM.

[164] D. Riboni, L. Pareschi, L. Radaelli, and C. Bettini. Is ontology-based activity recognition really effective? In *2011 IEEE International Conference on Pervasive Computing and Communications Workshops (PERCOM Workshops)*, pages 427–431, March 2011.

[165] Natalia Díaz Rodríguez, M. P. Cuéllar, Johan Lilius, and Miguel Delgado Calvo-Flores. A survey on ontologies for human behavior recognition. *ACM Comput. Surv.*, 46(4):43:1–43:33, March 2014.

[166] P. Salembier and J.R. Smith. MPEG-7 multimedia description schemes. *IEEE Trans. Circ. Syst. for Vid. Tech.*, 11(6):748–759, Jun 2001.

[167] Andrei Sambra, Henry Story, and Tim Berners-Lee. WebID 1.0. http://www.w3.org/2005/Incubator/webid/spec/identity/, 2014.

[168] Simone Santini and Amarnath Gupta. Disputatio on the use of ontologies in multimedia. In *Proceedings of the International Conference on Multimedia*, MM '10, pages 1723–1728, New York, NY, USA, 2010. ACM.

[169] S. Savarese, J. Winn, and A. Criminisi. Discriminative object class models of appearance and shape by correlatons. In *2006 IEEE Computer Society Conference on Computer Vision and Pattern Recognition*, volume 2, pages 2033–2040, 2006.

[170] Ansgar Scherp, Thomas Franz, Carsten Saathoff, and Steffen Staab. F – a model of events based on the foundational ontology Dolce+DnS ultralight. In *Proceedings of the Fifth International Conference on Knowledge Capture*, K-CAP '09, pages 137–144, New York, NY, USA, 2009. ACM.

[171] L. Schneider. Designing foundational ontologies – the object-centered high-level reference ontology OCHRE as a case study. In *Conceptual Modeling - ER 2003, 22nd International Conference on Conceptual Modeling*. Springer, 2003.

[172] Frank J. Seinstra, Jan-Mark Geusebroek, Dennis Koelma, Cees G. M. Snoek, Marcel Worring, and Arnold W. M. Smeulders. High-performance distributed video content analysis with parallel-horus. *IEEE MultiMedia*, 14(4):64–75, October 2007.

[173] Elhadi Shakshuki, Mohamed Kamel, and Hamada Ghenniwa. A multi-agent system architecture for information gathering. In *Proceedings of the 11th International Workshop on Database and Expert Systems Applications*, DEXA '00, pages 732–, Washington, DC, USA, 2000. IEEE Computer Society.

[174] Mehrnoush Shamsfard and Ahmad Abdollahzadeh Barforoush. Learning ontologies from natural language texts. *Int. J. Hum.-Comput. Stud.*, 60(1):17–63, January 2004.

[175] Ryan Shaw, Raphaël Troncy, and Lynda Hardman. LODE: Linking open descriptions of events. In *Proceedings of the 4th Asian Conference on the Semantic Web*, ASWC '09, pages 153–167, Berlin, Heidelberg, 2009. Springer-Verlag.

[176] J. Sivic, B. C. Russell, A. A. Efros, A. Zisserman, and W. T. Freeman. Discovering objects and their location in images. In *Tenth IEEE International Conference on Computer Vision, 2005. ICCV 2005.*, volume 1, pages 370–377 Vol. 1, Oct 2005.

[177] M. Slaney. Semantic-audio retrieval. In *2002 IEEE International Conference on Acoustics, Speech, and Signal Processing (ICASSP)*, volume 4, pages IV–4108–IV–4111, May 2002.

[178] A. W. M. Smeulders, M. Worring, S. Santini, A. Gupta, and R. Jain. Content-based image retrieval at the end of the early years. *IEEE Trans. Pattern Anal. and Mach. Intel.*, 22(12):1349–1380, Dec 2000.

[179] C. G. M. Snoek, B. Huurnink, L. Hollink, M. de Rijke, G. Schreiber, and M. Worring. Adding semantics to detectors for video retrieval. *IEEE Trans. Multimedia*, 9(5):975–986, Aug 2007.

[180] Cees G. M. Snoek, Marcel Worring, Jan C. van Gemert, Jan-Mark Geusebroek, and Arnold W. M. Smeulders. The challenge problem for automated detection of 101 semantic concepts in multimedia. In *Proceedings of the 14th Annual ACM International Conference on Multimedia*, MULTIMEDIA '06, pages 421–430, New York, NY, USA, 2006. ACM.

[181] Steffen Staab and Rudi Studer. *Handbook on Ontologies*. Springer Publishing Company, 2nd edition, 2009.

[182] Thomais Stasinopoulou, Lina Bountouri, Constantia Kakali, Irene Lourdi, Christos Papatheodorou, Martin Doerr, and Manolis Gergatsoulis. Ontology-based metadata integration in the cultural heritage domain. In *ICADL'07: Proceedings of the 10th International Conference on Asian Digital Libraries*, pages 165–175, Berlin, Heidelberg, 2007. Springer-Verlag.

[183] Jiang Su and Harry Zhang. Full Bayesian network classifiers. In *Proceedings of the 23rd International Conference on Machine Learning*, ICML '06, pages 897–904, New York, NY, USA, 2006. ACM.

[184] Erik B. Sudderth, Antonio Torralba, William T. Freeman, and Alan S. Willsky. Learning hierarchical models of scenes, objects, and parts. In *Proceedings of the Tenth IEEE International Conference on Computer Vision - Volume 2*, ICCV '05, pages 1331–1338, Washington, DC, USA, 2005. IEEE Computer Society.

[185] Afroza Sultana, Quazi Mainul Hasan, Ashis Kumer Biswas, Soumyava Das, Habibur Rahman, Chris Ding, and Chengkai Li. Infobox suggestion for Wikipedia entities. In *Proceedings of the 21st ACM International Conference on Information and Knowledge Management*, CIKM '12, pages 2307–2310, New York, NY, USA, 2012. ACM.

[186] S. Sural, Gang Qian, and S. Pramanik. Segmentation and histogram generation using the HSV color space for image retrieval. In *Image Processing. 2002. Proceedings. 2002 International Conference on*, volume 2, pages II–589–II–592 vol.2, 2002.

[187] Michael J. Swain and Dana H. Ballard. Color indexing. *Int. J. Comput. Vision*, 7(1):11–32, November 1991.

[188] Jie Tang, Ho-fung Leung, Qiong Luo, Dewei Chen, and Jibin Gong. Towards ontology learning from folksonomies. In *Proceedings of the 21st International Joint Conference on Artifical Intelligence*, IJCAI'09, pages 2089–2094, San Francisco, CA, USA, 2009. Morgan Kaufmann Publishers.

[189] Simon Tong and Edward Chang. Support vector machine active learning for image retrieval. In *Proceedings of the Ninth ACM International Conference on Multimedia*, MULTIMEDIA '01, pages 107–118, New York, NY, USA, 2001. ACM.

[190] Christopher Town. Ontology-driven bayesian networks for dynamic scene understanding. In *Proceedings of the 2004 Conference on Computer Vision and Pattern Recognition Workshop (CVPRW'04) Volume*

7 - *Volume 07*, CVPRW '04, pages 116–, Washington, DC, USA, 2004. IEEE Computer Society.

[191] Raphaël Troncy, Werner Bailer, Michael Hausenblas, Philip Hofmair, and Rudolf Schlatte. Enabling multimedia metadata interoperability by defining formal semantics of MPEG-7 profiles. In *Proceedings of the First International Conference on Semantic and Digital Media Technologies*, SAMT'06, pages 41–55, Berlin, Heidelberg, 2006. Springer-Verlag.

[192] Chrisa Tsinaraki, Panagiotis Polydoros, and Stavros Christodoulakis. Interoperability support between MPEG-7/21 and OWL in DS-MIRF. *IEEE Trans. on Knowl. and Data Eng.*, 19(2):219–232, February 2007.

[193] Chrisa Tsinaraki, Panagiotis Polydoros, Fotis Kazasis, and Stavros Christodoulakis. Ontology-based semantic indexing for MPEG-7 and TV-Anytime audiovisual content. *Multimedia Tools Appl.*, 26(3):299–325, August 2005.

[194] Peter D. Turney. Mining the web for synonyms: PMI-IR versus LSA on TOEFL. In *Proceedings of the 12th European Conference on Machine Learning*, EMCL '01, pages 491–502, London, UK, UK, 2001. Springer-Verlag.

[195] Vladimir N. Vapnik. *The Nature of Statistical Learning Theory*. Springer-Verlag New York, New York, NY, USA, 1995.

[196] P. Viola and M. Jones. Rapid object detection using a boosted cascade of simple features. In *Computer Vision and Pattern Recognition, 2001. CVPR 2001. Proceedings of the 2001 IEEE Computer Society Conference on*, volume 1, pages I–511–I–518 vol.1, 2001.

[197] Akrivi Vlachou, Christos Doulkeridis, and Yannis Kotidis. Metric-based similarity search in unstructured peer-to-peer systems. In Abdelkader Hameurlain, Josef Küng, and Roland Wagner, editors, *Transactions on Large-Scale Data- and Knowledge-Centered Systems V*, pages 28–48. Springer-Verlag, Berlin, Heidelberg, 2012.

[198] Julia Vogel and Bernt Schiele. Semantic modeling of natural scenes for content-based image retrieval. *Int. J. Comput. Vision*, 72(2):133–157, April 2007.

[199] D. Vogiatzis, D. Pierrakos, G. Paliouras, S. Jenkyn-Jones, and B. J. H. H. A. Possen. Expert and community based style advice. *Expert Syst. Appl.*, 39(12):10647–10655, September 2012.

[200] Stefanos Vrochidis, Charalampos Doulaverakis, Anastasios Gounaris, Evangelia Nidelkou, Lambros Makris, and Ioannis Kompatsiaris. A hybrid ontology and visual-based retrieval model for cultural heritage multimedia collections. *Int. J. Metadata, Sem. and Ontologies*, 3(3):167–182, 2008.

[201] RDF schema 1.1. http://www.w3.org/TR/rdf-schema/.

[202] HTML 4.0 Specification. http://www.w3.org/TR/1998/REC-html40-19980424/, 1998.

[203] Extensible markup language (XML) 1.0 (fifth edition). http://www.w3.org/TR/xml/, 2008.

[204] Notation3 (N3): A readable RDF syntax. http://www.w3.org/TeamSubmission/n3/, 2011.

[205] RIF overview (second edition). http://www.w3.org/TR/rif-overview/, 2013.

[206] SPARQL query language for RDF. http://www.w3.org/TR/rdf-sparql-query/, 2013.

[207] Cross-origin resource sharing. http://www.w3.org/TR/cors/, 2014.

[208] HTTP - Hypertext Transfer Protocol. http://www.w3.org/Protocols/, 2014.

[209] RDF 1.1 concepts and abstract syntax. http://www.w3.org/TR/2014/REC-rdf11-concepts-20140225/, 2014.

[210] RDFa core 1.1 - third edition. http://www.w3.org/TR/rdfa-core/, 2015.

[211] World Wide Web Consortium (W3C). Web ontology language. http://www.w3.org/TR/owl2-overview/.

[212] Jingya Wang, Mohammed Korayem, and David J. Crandall. Observing the natural world with flickr. In *Proceedings of the 2013 IEEE International Conference on Computer Vision Workshops*, ICCVW '13, pages 452–459, Washington, DC, USA, 2013. IEEE Computer Society.

[213] Xiaofeng Wang, Gang Li, Guang Jiang, and Zhongzhi Shi. Semantic trajectory-based event detection and event pattern mining. *Knowledge and Information Systems*, 37(2):305–329, 2013.

[214] Sujal Subhash Wattamwar and Hiranmay Ghosh. Spatio-temporal query for multimedia databases. In *Proceedings of the 2nd ACM Workshop on Multimedia Semantics*, MS '08, pages 48–55, New York, NY, USA, 2008. ACM.

[215] Cyc: From Wikipedia, the free encyclopedia. http://en.wikipedia.org/wiki/Cyc (last accessed: April 2015).

[216] DBpedia: From Wikipedia, the free encyclopedia. http://en.wikipedia.org/wiki/Cyc (last accessed: April 2015).

[217] Female body shape: From Wikipedia, the free encyclopedia. http://en.wikipedia.org/wiki/Female_body_shape (last accessed: April 2015).

[218] Erling Wold, Thom Blum, Douglas Keislar, and James Wheaton. Content-based classification, search, and retrieval of audio. *IEEE MultiMedia*, 3(3):27–36, September 1996.

[219] Wilson Wong, Wei Liu, and Mohammed Bennamoun. Tree-traversing ant algorithm for term clustering based on featureless similarities. *Data Mining and Knowledge Discovery*, 15(3):349–381, 2007.

[220] Wilson Wong, Wei Liu, and Mohammed Bennamoun. Ontology learning from text: A look back and into the future. *ACM Comput. Surv.*, 44(4):20:1–20:36, September 2012.

[221] Pengcheng Wu, Steven C.H. Hoi, Hao Xia, Peilin Zhao, Dayong Wang, and Chunyan Miao. Online multimodal deep similarity learning with application to image retrieval. In *Proceedings of the 21st ACM International Conference on Multimedia*, MM '13, pages 153–162, New York, NY, USA, 2013. ACM.

[222] Xiao Wu, Yi-Jie Lu, Qiang Peng, and Chong-Wah Ngo. Mining event structures from web videos. *IEEE MultiMedia*, 18(1):38–51, January 2011.

[223] Rong Xiao, Long Zhu, and Hong-Jiang Zhang. Boosting chain learning for object detection. In *Proceedings of the Ninth IEEE International Conference on Computer Vision - Volume 2*, ICCV '03, pages 709–, Washington, DC, USA, 2003. IEEE Computer Society.

[224] Hongtao Xu, Xiangdong Zhou, Mei Wang, Yu Xiang, and Baile Shi. Exploring flickr's related tags for semantic annotation of web images. In *Proceedings of the ACM International Conference on Image and Video Retrieval*, CIVR '09, pages 46:1–46:8, New York, NY, USA, 2009. ACM.

[225] Rong Yan, Marc-Olivier Fleury, Michele Merler, Apostol Natsev, and John R. Smith. Large-scale multimedia semantic concept modeling using robust subspace bagging and mapreduce. In *Proceedings of the First ACM Workshop on Large-scale Multimedia Retrieval and Mining*, LS-MMRM '09, pages 35–42, New York, NY, USA, 2009. ACM.

[226] Rong Yan and A. Hauptmann. Multi-class active learning for video semantic feature extraction. In *2004 IEEE International Conference on Multimedia and Expo, 2004. ICME '04*, volume 1, pages 69–72 Vol.1, June 2004.

[227] Rong Yan and M. Naphade. Semi-supervised cross feature learning for semantic concept detection in videos. In *IEEE Computer Society*

Conference on Computer Vision and Pattern Recognition, 2005. CVPR 2005, volume 1, pages 657–663 vol. 1, June 2005.

[228] Cheng Yang. Peer-to-peer architecture for content-based music retrieval on acoustic data. In *Proceedings of the 12th International Conference on World Wide Web*, WWW '03, pages 376–383, New York, NY, USA, 2003. ACM.

[229] Chengyong Yang, Erliang Zeng, Tao Li, and G. Narasimhan. Clustering genes using gene expression and text literature data. In *2005 IEEE Computational Systems Bioinformatics Conference, 2005. Proceedings*, pages 329–340, Aug 2005.

[230] Kuiyuan Yang, Meng Wang, Xian-Sheng Hua, and Hong-Jiang Zhang. Tag-based social image search: Toward relevant and diverse results. In Steven C. H. Hoi, Jiebo Luo, Susanne Boll, Dong Xu, Rong Jin, and Irwin King, editors, *Social Media Modeling and Computing*, pages 25–45. Springer London, 2011.

[231] Bangpeng Yao, Aditya Khosla, and Li Fei-Fei. Combining randomization and discrimination for fine-grained image categorization. In *2011 IEEE Conference on Computer Vision and Pattern Recognition (CVPR)*, pages 1577–1584. IEEE, 2011.

[232] Chi Zhang, Feifei Li, and Jeffrey Jestes. Efficient parallel knn joins for large data in mapreduce. In *Proceedings of the 15th International Conference on Extending Database Technology*, EDBT '12, pages 38–49, New York, NY, USA, 2012. ACM.

[233] Tong Zhang and C.-C.J. Kuo. Audio content analysis for online audio-visual data segmentation and classification. *IEEE Trans. Speech and Audio Processing*, 9(4):441–457, May 2001.

[234] Zijian Zheng and Geoffrey I. Webb. Lazy learning of bayesian rules. *Mach. Learn.*, 41(1):53–84, October 2000.

[235] Ning Zhong, JianHua Ma, RunHe Huang, JiMing Liu, YiYu Yao, YaoXue Zhang, and JianHui Chen. Research challenges and perspectives on wisdom web of things (w2t). *The Journal of Supercomputing*, 64(3):862–882, 2013.

[236] Lina Zhou. Ontology learning: State of the art and open issues. *Inf. Technol. and Management*, 8(3):241–252, September 2007.

[237] Wenwu Zhu, Chong Luo, Jianfeng Wang, and Shipeng Li. Multimedia cloud computing. *IEEE Signal Processing Magazine*, 28(3):59–69, May 2011.

[238] Kuangtian Zhufeng, Lei Wang, Hanli Wang, Yun Shen, Wei Wang, and Cheng Cheng. Large-scale multimedia data mining using mapreduce framework. In *Proceedings of the 2012 IEEE 4th International Conference on Cloud Computing Technology and Science (CloudCom)*, CLOUDCOM '12, pages 287–292, Washington, DC, USA, 2012. IEEE Computer Society.

Index